Home Plumbing Mastery

*Your Essential Guide to Understanding and Tackling
Plumbing Challenges with Confidence*

CONNOR CARTER

Connor Carter

Table of Contents

Connor Carter

Introduction:

Plumbing. For most of us, it's one of those things that we barely notice—until something goes wrong. A dripping faucet. A clogged toilet. A leaky pipe. Then suddenly, we're faced with the mysteries of shut-off valves, drain traps, and all sorts of plumbing arcana. That's where this book comes in. Consider it your guide to the wonderful yet complex world of home plumbing. With a roll-up-your-sleeves, let's-dive-in approach, we'll explore all the essential systems and components that make up your home's plumbing. Water supply lines. Drain and vent piping. Fixtures and appliances like sinks, toilets, and water heaters. I'll teach you how it all works together to bring water in and send waste out. We'll cover the basic tools and safety tips you need to become an amateur home plumber. Before you know it, you'll be talking PVC and PEX like a pro. We'll troubleshoot common problems like leaks, clogs, and loss of water pressure. And we'll fix them ourselves with easy-to-follow repair techniques. From unclogging drains to replacing faucet washers, we've got this covered. What if we could turn household plumbing from a dry, technical topic into a fun adventure? A journey of discovery, with hard-won lessons and muddy misadventures? An epic tale of you against the pipes? Well, friend, that's exactly what we'll do here.

Let's make plumbing whimsical again. To liven things up, we'll start each chapter with amusing anecdotes and random facts about plumbing history. (Did you know some of the first indoor flush toilets in the 1800s had to be manually pumped by hand?) We'll get to know some of the giants of the plumbing world, like Thomas Crapper, inventor of the ballcock valve. And we'll even bust some persistent plumbing myths. (No, toilets do NOT flush in a different direction in the Southern Hemisphere.

This book has two goals: one practical, one philosophical. First, by the time you finish, you'll have the key skills and knowledge to handle basic repairs yourself instead of calling an expensive pro. But also, I want to change your attitude toward plumbing. To get you excited about the mechanics of moving water around your home. To show you that plumbing has beauty and meaning far beyond just keeping your basement dry. With the right mindset and a few tools, you can become the master of your plumbing domain. You'll gain confidence to tackle repairs and face problems head-on rather than ignoring them. The same leak you used to curse will now inspire you to roll up your sleeves and get to work. Too often we take plumbing for granted, forgetting the comfort it brings to our daily lives. This book is a celebration of the humble pipe wrench and the quiet nobility of the home plumber. Join me on this squishy, sludgy ride through the inner workings of your house. Modern "plumbing" refers to a wide array of supply pipes, drains, vents, valves, fittings and fixtures. While homeowners will occasionally encounter minor leaks or clogs requiring basic repairs, professional plumbers have the expertise to handle installation, complex diagnostics, maintenance and large-scale renovation projects. They understand critical concepts like water pressure, drainage venting, material compatibility, building codes and safety standards.

This book will give you a new appreciation for plumbing by walking you through these key principles, systems, tools and skills. You'll learn how your home's water enters from the main line, gets distributed through an array of pipes, and empties into drains and sewers. We'll cover fixtures like sinks, toilets, tubs, faucets, water heaters and appliances connecting to this infrastructure. You'll learn enough to become an educated homeowner who can confidently tackle minor repairs and maintenance.

So let your fascination with plumbing begin! This knowledge will enrich your home ownership experience and allow you to converse intelligently with plumbers when major issues arise. Understanding the concepts behind proper plumbing design and function will help you make upgrade decisions that maximize safety, efficiency and sustainability. Consider this book an investment—both in your peace of mind and in your home's long-term maintenance. The next time your pipes leak or your faucet drips, you'll respond with curiosity rather than panic. Let's dive in!

Chapter 1

The Wonderful World of Plumbing

Plumbing has been an essential part of civilized human life for thousands of years, evolving from basic waste drainage systems in ancient Mesopotamia to the sophisticated piping networks we rely on today. While early plumbing methods did not involve pipes, drainage systems were already being implemented in royal palaces around 2500 BCE to transport waste away from cities and into covered cesspools. The Romans took plumbing further by developing aqueducts, sewers, lead pipes and flush toilets to supply water and remove waste on a grand scale. After the fall of Rome, however, plumbing technology stagnated for many centuries as most ordinary people continued to rely on wells, rain barrels and chamber pots into the Middle Ages. It wasn't until around 1200 CE that wooden hollowed logs started being used more commonly as primitive pipes. By the 1500s, pipes made from lead, tin and copper were being manufactured for the plumbing systems of wealthy homes and large public buildings. But it still remained an arduous chore to fetch water and empty chamber pots for most people. The spread of indoor plumbing truly accelerated in the mid-1800s thanks to several key innovations. The invention of the modern flush toilet with an integrated flushing mechanism in 1829 was a major breakthrough, followed by the vulcanization of rubber in 1850 for flexible hoses and gaskets. Louis Pasteur's germ theory in 1861 made people aware of the detrimental health hazards of poor sanitation, creating demand for better waste removal. On the heels of these discoveries, the rubber flush valve ballcock was patented by Thomas Crapper in 1883, standard pipe sizes were established in 1885, and the crucial S-trap was invented by Thomas Campbell in 1889.

Home Plumbing Mastery

With these advancements in place, indoor plumbing started becoming mainstream in wealthier urban households by the early 1900s, though retrofitting systems into existing homes remained labor-intensive plumbing guesswork. Thankfully by the 1920s, the integration of plumbing during building construction instead of after the fact, along with standardized building codes and materials like galvanized steel pipes, made the installation process more consistent and accessible. The pace of innovation sped up through the Space Age, with the development of PVC plastic pipes in the 1940s and cross-linked polyethylene (PEX) tubing in the 1950s, both game-changing materials that enabled faster and cheaper plumbing setup. Today, we take for granted conveniences like instant hot and cold running water and waste that disappears with the flush of a lever—unimaginable luxuries for most of human civilization made possible by over 5,000 years of plumbing ingenuity. When turning a faucet or jiggling a toilet handle, it helps to reflect on the ancient Roman aqueducts, pioneering Victorian sanitation reformers, and polymer scientists who laid the groundwork for the everyday magic of plumbing.

Beyond just convenience, plumbing has been one of the most significant public health interventions in history. Before modern sanitation, cities were breeding grounds for diseases like cholera and typhoid spread by contaminated water and inadequate waste removal. Epidemics frequently decimated urban populations. The "sanitary revolution" that accompanied the rise of clean piped water and sewer systems from the mid-1800s onward dramatically increased life expectancy and reduced infant mortality by preventing these deadly yet preventable infections. This transformation in public health underscored a new attitude that human engineering could conquer problems once fatalistically accepted as inevitable.

Progressive cities became proactive engines of hygiene through plumbing plus reforms in housing, food safety, sanitation, and garbage removal. While global access to safe water and sanitation remains an urgent issue today, the plumbing innovations made over the past 150 years have vastly improved public health outcomes overall. For those fortunate enough to have plumbing, its quiet life-saving role is easy to overlook as we take for granted what was once inconceivable luxury. Beyond just health, indoor plumbing also immensely improved quality of life by eliminating the daily drudgery of hauling water and cleaning chamber pots.

By enabling urban density with fewer health drawbacks and making buildings habitable year-round, plumbing protections against the elements helped transform homes into truly livable private spaces. The next time a drain empties smoothly, or shower runs hot, take a moment to appreciate how pipes inside the walls represent public health triumphing over disease, comfort defeating drudgery, and scientific progress enhancing life. Home plumbing is more than mere convenience—it is a quiet hero that has profoundly pushed civilization forward.

While plumbing can feel like a solitary struggle when tackling a leaky basement faucet, there is an entire community of fellow DIY plumbers out there connected by a shared passion. Online forums make it easy to exchange tips, ask questions, and share project stories and blunders with appreciative audiences who just get it. This virtual knowledge-sharing helps cultivate expertise as newcomers learn from seasoned plumbers passing down skills that otherwise take years to develop. With crowd-sourced wisdom, solutions can be found even for the most stubborn plumbing riddles. When a project goes sideways, you can post a picture and get real-time troubleshooting advice to quickly get unstuck. A spirit of camaraderie develops over shared mishaps and triumphs as members become familiar faces. Experienced mentors happily pay it forward to anxious newbies. And the spaces get delightfully weird with heated debates on topics like toilet paper orientation and bidet etiquette alongside technical know-how and inside jokes. While the internet forms connections, the local plumbing community thrives through trade schools, union apprenticeships, supply shops, and industry events. Make use of online resources, but also seek out nearby enthusiasts to learn collectively. Understand that no plumber is an island! This far-flung yet tight-knit community is unified by a core belief that plumbing is not just utilitarian necessity, but a craft capable of beauty and mastery with always more to learn.

Home Plumbing Mastery

While grounded in practical science, plumbing has also spawned many peculiar superstitions through the years. According to maritime tradition, beginning a voyage on a Friday invites bad luck, reinforced by frequent plumbing failures on ships that day when amateur plumbers did repairs. Ancient Babylonian texts warn against using a toilet at night lest demons enter through the pipes and possess your body. In parts of Europe, many still believe doing laundry on New Year's Day may wash away a family member in the coming year. Feng shui practitioners advise against positioning a toilet near the front door or kitchen for its supposedly disruptive energy flow. And in some cultures, using the left hand instead of the right to unclog drains is considered terrible luck as it is associated with impurity. While we can laugh off these myths today, they signify the aura of mystery pipes once held.

In addition to superstitions, the history of plumbing contains many fascinating facts and colorful characters. The word "plumber" derives from the Latin for lead, plumbum, as early plumbers worked primarily with lead pipes. Plumbing first appeared in English in the 1400s. Women were banned from medieval plumbing guilds and did not begin entering the trade until the 1940s. Plumbers adopted the Roman god Janus, often depicted with two faces, as a patron saint of sorts. The famous nursery rhyme "Sing a Song of Sixpence" is thought to reference the blackbirds that got stuck and died in London's filthy early plumbing. Mario and Luigi of Super Mario Bros were originally cast as plumbers. And plumbers can claim patron saints like Saint Benedict who once miraculously caused a fountain to appear. So while the drain-clearing, pipe-fixing stereotype persists, the legacy of plumbing is rich in odd trivia and fascinating facts proving plumbers are anything but boring!

The key breakthroughs that evolved plumbing into its modern form also make for an exciting history. Mesopotamian drainage and sewers, Roman aqueducts, Thomas Crapper's flush toilet, galvanized steel pipes, PVC plastic, and push-fit fittings each improved health, accessibility, reliability and ease of installation over thousands of years. Each generation of clever plumbers found new ways to transport water safely, adding up to the convenient systems enjoyed today. When we understand how long and challenging that journey has been, it becomes easier to appreciate the everyday magic of turning a faucet or flushing waste without a second thought. We owe immense gratitude to the innovators across cultures and eras who made plumbing a flexible, adaptable technology that has so enormously impacted civilization.

Plumbing practices and customs around the world showcase human ingenuity in adapting to local conditions. In rural parts of India lacking piping, the traditional practice of falanga involves carrying water long distances by metal containers on foot. Dry composting toilets are growing in popularity in water-scarce regions, separating liquid and solid waste through evaporation and decomposition instead of water-intensive flushing.

Solar water disinfection utilizes UV rays to purify drinking water in areas without infrastructure by simply placing water bottles in the sun. The tropical Asian custom of mandi replaces toilet paper with cleansing after toileting using water scooped from a tabo, considered more hygienic. Japanese high-tech toilet innovation focuses on luxury enhancements like heated seats, massage jets, and sound effects. Ancient underground qanat aqueducts up to 300 km long sustain settlements through arid Middle Eastern climates to this day. While modern plumbing homogenizes things, echoes of these unique regional practices persist worldwide.

Chapter 2

Basic Tools for the Home Plumber

Any home plumber worth their salt knows a good set of wrenches and pliers can fix most problems. While advanced gadgets have their place, there's no substitute for these classic manual tools. Let's cover the essential wrenches and pliers to include in your tool kit.

First, you need both standard wrenches and adjustable wrenches. Standard wrenches come in specific sizes like 1/4", 5/16", etc. to fit different fasteners. Every home plumber should have a set including common screw and bolt sizes. Adjustable wrenches like crescent wrenches can adjust to grip various widths. Get a 10" and a 6" adjustable wrench to start.

Pipe wrenches, with their serrated jaws, are a must for gripping round fixtures and pipes during assembly or disassembly. A 14" pipe wrench will handle most residential plumbing tasks. Mini pipe wrenches are handy for tight spots.

Don't forget the humble monkey wrench. It adjusts like a crescent wrench, but its two-piece jaw design provides more gripping power on pipes and nuts. They're useful in a pinch when a pipe wrench isn't available. Speaking of grip, channel lock pliers should be high on your list. Their adjustable pinned joint allows you to really clamp down on plumbing fixtures and fasteners. The serrated jaws grip slick surfaces. Channel locks come in different lengths - a 10" and 6" pair will be useful. Needle nose pliers are essential for precision work in tight spaces. Their tapered, pointed jaws can grip small fittings, fasteners, and wires where fingers can't reach. They also help bend and reposition small pipes. Lastly, don't forget safety. Wear goggles and gloves when force is needed to loosen a stubborn bolt or pipe—wrenches can slip. Also choose quality tools—cheap wrenches that break or slip can lead to injury.

Home Plumbing Mastery

Invest in the best standard, adjustable, pipe, and channel lock wrenches you can afford. And keep them clean and maintained so they continue gripping properly. With this classic wrench and plier lineup in your tool kit, you'll be ready to handle all sorts of plumbing challenges!

When it comes to residential plumbing projects, you'll inevitably need to cut and join pipes, tubes, and fittings. Stock your tool kit with these essential cutting and joining tools. For basic pipe cutting, a standard hacksaw with a metal cutting blade works for copper, plastic, and thin steel pipes. Hacksaw blades come in different tooth patterns—choose one with fine teeth for smoother cuts. A miter box guides angled cuts. Moving up, a pipe cutter tool makes perfectly straight cuts on copper pipe up to 1 inch diameter. Just crank the cutter wheel around the pipe's circumference to slice through. Clean up any burrs with a file or reaming tool.

For larger steel and plastic pipes, nothing beats a power reciprocating saw. An adjustable blade lets you cut pipes up to 6 inches diameter, and reciprocating saws can also demolish old pipe if needed. Protective goggles are a must for power saws.

PVC and ABS plastic pipes require special treatment. A PVC pipe shear or shear blade on a power saw makes straight, burr-free cuts in plastic drain piping so joints fit cleanly. Detailed measurements matter with plastic pipe.

Now let's discuss options for joining cut pipe ends. For copper pipes, soldering still reigns supreme. Apply flux paste, insert a copper joiner fitting, and heat with a propane torch while applying solder. Sand cloth preps joints.

Compression fittings provide a simpler copper pipe joint solution. Just insert the pipe end into the fitting and tighten the nut to form a sealed joint. No torch required! Great for the DIY novice. For PVC and ABS pipes, solvent cement joins and permanently bonds plastic fittings and joints. PVC primer preps the surfaces first—then apply solvent cement and quickly join parts. Clamps temporarily hold joints while cement dries. Steel and plastic threading tools cut male threads on pipe ends to fit female threaded fittings. Dies do interior threads while taps cut exterior threads. A pipe vise secures pipes while threading.

Finally, protect your eyes and hands while cutting and joining. And practice techniques first! Patience and care pays off in smooth, lasting plumbing joints. With this diverse lineup of cutting and joining tools, you'll have the power to fabricate leak-free plumbing repairs and new installations no matter the pipe type or size. Your materials don't stand a chance.

Plumbing systems require vigilance to spot issues before they turn into bigger problems. A home plumber's cleaning and inspection toolkit makes regularly monitoring pipes, fixtures, and drains a breeze. First, a mirror on an extensible pole is ideal for peeking under sinks or into other hard-to-reach spots. Check drainpipes and garbage disposals for leaks, obstructions, or deterioration. An LED adds light. For drains, a Zip-It plastic cleaning tool clears hair clogs when inserted in a drain opening and pulled back. For deeper drain cleaning, plastic drain snakes of different lengths attach to drill motors to grind through serious clogs.

Don't forget about your household drain nemesis—hair, grease, and grime. Bottled liquid drain cleaners break down organic gunk and freshen up drains every month or so. Use a funnel for accuracy. Accessing pipes often means removing drain trap covers beneath sinks and tubs. A Trap Wrench fits the curve and bolts of most tub and sink traps for quick removal and replacement during inspections.

Pipes and fixtures need scrubbing too. Scotch-Brite scouring pads can be attached to drills to polish away corrosion and mineral deposits inside pipes exposed by cut-outs or detachments. For leaks, a moisture meter can detect elevated dampness in drywall and ceiling material before visible water damage appears. This lets you pinpoint origin points. An infrared thermometer also identifies temperature differences.

Got a sewer gas smell? A combustible gas detector alerts you to dangerous hydrogen sulfide or methane buildup from sewer line leaks. Sniff out hazards before they become bigger issues.

Finally, a mechanical listening tool amplifies noises in pipes and fixtures to aid diagnosis. Strange gurgles or hisses indicate flow problems. Knocking could signal loose piping. Get familiar with pipe sounds. With this cleaning and inspection toolkit, you can keep household plumbing in tiptop shape.

Home Plumbing Mastery

Plumbing sometimes involves contending with hazardous situations like sharp tools, electricity, and leaking water. Don't take safety lightly—a bit of protective gear goes a long way in keeping you injury-free. First, eye protection is a must when soldering, using power tools, or cleaning pipes with chemicals. Safety glasses block particles and debris. For tasks with flying sparks or grit, choose goggles or a face shield. Ear protection like earmuffs or earplugs saves your hearing. Running pump motors, power saws, and grinding tools get loud enough to damage eardrums over time. It's important to muffle the noise to preserve your hearing.

Respiratory safety matters too. Dust masks filter out airborne particles when sawing pipe or sanding joints. Respirators with organic vapor cartridges protect against fumes from glues, solvents, and cleaners. Hand safety should be top of mind at all times . Leather or canvas work gloves provide abrasion protection and grip when handling pipes and tools. Insulated rubber gloves shield hands from electrical shocks.

Speaking of electricity, insulated mats protect against ground faults when maintaining conductive appliances like water heaters. Never touch electrical components without isolating power first. Safety first!

Knee pads cushion impact when crawling on floors installing drain lines or under sinks. Comfortable kneepads mean less pain and ability to work longer. Proper attire like long sleeves and pants is crucial when soldering or using power tools. Gear can catch on spinning equipment. Durable boots with toe protection fend off dropped tools and heavy items.

Lastly, keep a well-stocked first-aid kit nearby for minor injuries. Plumbing presents opportunities for cuts, scrapes, burns, and pinched fingers. Disinfect and dress wounds promptly. Safety costs a small fraction compared to an injury. Use common sense and invest in quality protective equipment appropriate for each plumbing task. It takes little effort to work safely.

Beyond essential wrenches and cutters, some specialty gadgets and accessories make plumbing projects easier. Let's cover handy instruments that simplify layout, measurement, assembly, and repairs. First, a thin, flexible auger snake threads into pipes, elbows, and drains to clear clogs in tight spots wrenches can't reach. Hand cranks gently work through obstructions. Protect pipes by starting with plastic snakes before resorting to steel.

Leaky pipe joints call for epoxy putty sticks. Just knead, apply, and smooth putty around leaks for permanent seals safe for potable water. Great temporary fix for pinholes and cracks. A spring-loaded tube cutter makes clean, straight slices in flexible tubing like hoses and PEX lines up to 1 inch diameter. Just squeeze the handle and rotate around tubing. No saw needed!

Protecting pipe threads during assembly prevents leaks. Pipe thread seal tape wraps around male threaded fittings to seal threads. Paste sealant works too. Apply to thread peaks for leak-free joints. Locating studs to anchor fixtures often involves trial and error. Save time with an electronic stud finder, which detects wood studs behind drywall via density changes. Take the guesswork out of mounting. When soldering copper pipes, flux brushes apply acid flux to joint areas, while wire joint scrubbers buff away oxidation so solder adheres well.

A properly equipped workbench helps you tackle plumbing projects efficiently and safely. Here are some essential items for an ideal plumbing workstation.

A roomy, sturdy workbench provides a stable surface for tasks like soldering, gluing, and assembling fixtures. Choose a height comfortable for you. Add locking casters for mobility. A vise with padded jaws secures pipes and fittings for assembly and modifications. It should mount securely to the edge of the workbench and open wide enough for large pipe diameters. Mount a pegboard above the bench to organize tools within easy reach. Outline tool silhouettes so you can tell if any are missing at a glance. Good lighting is crucial for delicate work. Use an adjustable desk lamp and overhead shop lights. High lumen LED bulbs provide crisp light without glare. Keep a fire extinguisher and first aid kit nearby for safety along with eye wash solution. Also mount an accessible trash can for debris and scrap material. Designate drawers for organized plumbing parts like O-rings, washers, and valves. Label them clearly for quick identification without sorting drawers. With a clean, well-lit, and efficiently organized workbench, plumbing projects feel controlled rather than chaotic. Finally, a folding work seat lets you tackle tasks comfortably without contorting or kneeling. A sturdy adjustable work seat makes projects more ergonomic and enables longer work sessions.

Having the right portable toolbox setup makes house calls and remote plumbing jobs smooth. Here are recommended tools to keep on hand for portability:

Adjustable wrenches in 10" and 6" sizes
Slip joint pliers with curved jaws
Set of socket wrenches with common metric and imperial sizes
Assorted standard and Phillips head screwdrivers
Set of hex keys for appliance fasteners
Flat and straight edge putty knives for gasket work
Claw hammer for light demolition and fastener removal
Utility knife with extra blades
Tape measure at least 10 feet long
Headlamp for hands free illumination
Safety glasses and work gloves
Knee pads for crawl spaces
Tailor your portable toolkit to the types of plumbing you do. The goal is having essentials handy, so you don't waste time retrieving tools.

In addition to standard tools, some obscure specialty implements make specific plumbing tasks much easier:
Offset flaring tools for precisely sizing copper tubing ends
Ratchet pipe cutters for quick, clean plastic pipe cutting
Flexible auger for snaking laundry drain lines
Moen cartridge puller for removing faucet valves
Water pressure gauge kit for diagnosing supply issues
Infrared thermometer for finding leaks behind walls
Researching the ideal specialty tools for your needs helps you work more efficiently and professionally. But don't overbuy—purchase purpose-made tools only as needed.

Cleaning and organizing tools makes plumbing work more efficient and extends tool life. Follow these tips:
Wipe down tools after use to prevent residue buildup
Lubricate adjustable wrench and plier joints
Dry tools if exposed to water to prevent corrosion
Sharpen cutting blades when dull
Coil hoses, cords, and cables neatly to prevent kinks
Return tools to labeled storage locations so they can be found quickly
Hang commonly used tools on workshop pegboards for quick access
Inventory kits annually and replace missing or damaged items
Proper tool care and organization prevents downtime wasted searching for tools or dealing with damaged equipment during a project. Develop good maintenance habits.

With these convenient accessories, common plumbing jobs stay quick, clean, and hassle-free. Assemble your tool kit with the essential gadgets and helpers that streamline your specific projects. Work smarter, not harder!

Modern supply plumbing uses a variety of pipe materials, each with advantages and uses. Understanding different properties helps select the right supply lines.

Copper - Long the gold standard for supply lines, copper is durable, corrosion-resistant, and easily soldered. However, inflexible copper pipes can be tricky to install in tight spaces.

PEX - Flexible cross-linked polyethylene tubing doesn't corrode and resists scale buildup. PEX is easy to install around corners, but it cannot withstand as much heat as copper.

PVC - Durable, inexpensive plastic PVC is most common for modern drain lines. It is easy to cut and glue together. However, it can become brittle with exposure to chemicals.

Lead - Once common in plumbing, health concerns mean lead is no longer used for new supply lines. Lead solder has also been phased out. Existing lead lines should be replaced.

Cast Iron - Traditional cast iron pipes are long-lasting but heavy and rigid. Older homes still rely on cast iron drains that require special couplings when joining to other materials.

ABS Plastic - Lightweight ABS plastic pipes are easy to work for DIYers. But ABS is not as strong as PVC and more likely to shatter if frozen. It should not be used for main drains.

Clay - Historically used for drains, salt glazed clay pipes are extremely durable but rigid. Roots can penetrate clay pipe joints over time. Replacement parts are hard to find.

Concrete - Concrete pipes are strong but very heavy, only feasible for major municipal projects. However, smaller concrete pipes are sometimes used in home septic applications.

Understanding these common drainpipe materials allows for proper repair, coupling, and replacement as aging drain lines fail. Choose materials that balance durability, cost, weight, and ease of installation.

Creating a map of your home's plumbing system also helps immensely with preventative maintenance and simplifies repairs when issues arise. Here are some tips for DIY plumbing mapping:

Home Plumbing Mastery

Trace supply and drain lines through the house, basement, and exterior. Label fixtures connected to each line.

Note pipe materials and sizes wherever visible. Look for size information printed on pipes.

Identify the locations of shutoff valves, cleanouts, and water meter. These are crucial during emergencies.

Draw a schematic showing how lines are interconnected. Indicate direction of water flow with arrows.

Mark the areas lines are concealed, like under slabs or in walls. This informs future access if repairs are needed.

Take photos of key components to supplement your line drawings and notes.

Talk to any household members knowledgeable about the systems. Long-term residents may recall helpful details.

Search for any original home plumbing blueprints, even if outdated. These provide insights into the overall system.

Update your map anytime modifications are made. Add dates to show plumbing history.

A detailed map doesn't need artistic skill, just patience and curiosity. Take time deciphering your home's hidden plumbing secrets—it will pay back dividends through easier maintenance and repairs.

Chapter 3

Understanding Your Home's Plumbing System

The hidden network bringing fresh water into your home is the water supply system. Understanding how it works helps diagnose issues and make repairs. Let's overview the components supplying water for bathing, cleaning, and drinking. It starts with the main water supply line, a large pipe connecting your home to the main municipal water system or a private well. This feed line branches into two main circuits—the cold water supply and hot water supply.

The cold side powers fixtures like sinks, tubs, and toilets directly with untreated water. Copper, PEX, or galvanized steel pipes form an organized branching network running water to each fixture.

Shut-off valves placed strategically allow isolating sections of the cold supply for maintenance. You should know where your main shut-off is located, in case of emergencies!

For hot water, the cold supply feeds into a hot water heater tank. Heating elements or a gas burner warms the water, which then flows into the hot water supply circuit. This loop runs hot water to all fixtures needing it. The hot water loop has its own shut-off valve too. Insulated pipes prevent heat loss and keep hot water energized on its way to taps and showers. Proper insulation saves energy and keeps water usable.

Understanding this basic supply anatomy allows you to systematically track down leaks and issues. Follow the supply chain backward from fixture to main when troubleshooting. And take time to familiarize yourself with shut offs so you can act fast in an emergency.

Of course, surface details vary house to house. But the underlying premise of a branched supply remains the same. Trace your supply lines, label components, and get to know your system's quirks. Then tackling any water supply headache becomes much less daunting.

While supply pipes bring in fresh water, drains and vents whisk wastewater away. Understanding how drainage systems work helps keep them flowing freely. It all starts at fixtures like sinks, tubs, and toilets. Tailpieces, tub drains, and toilet flanges connect to wider drainpipes, often made of PVC or cast iron. Angled fittings route water downward through vertical stacks. Vent pipes run up alongside drains to allow air flow. This prevents vacuum pressure and siphoning problems. Proper venting lets drains completely empty.

Drainpipes converge as they run downward. Bathroom groups join together, then connect to larger kitchen and laundry drains. Eventually all drainage empties into a main sewer line exiting your home. Whether municipal sewer or private septic system, the final runoff flows to a central waste treatment plant. Careful venting and slope allow waste to drain fully using only gravity.

Of course, clogs happen! That's where cleaning and augering tools come in handy. Regular upkeep is key to keep drains clear and functioning. A clog anywhere can block the entire system. Pay attention to gurgles or slow drains and address issues early before major blockages develop. Awareness of the overall drain architecture helps isolate problem spots.

Like supply lines, labeling your drains, stacks, and vents makes repairs easier. A detailed diagram helps track down clogs and leaks affecting a specific fixture or room. You want to know your drainage anatomy! So in summary, drainage starts small then converges downward until exiting the home, smooth vented flow prevents clogs and backups, and attentiveness keeps waste moving smoothly without unhealthy stagnation.

Fixtures and appliances are the interface between you and your home's plumbing system. Understanding how these everyday components work helps you use and maintain them properly. The most common fixtures are sinks, toilets, showers, and bathtubs. Sinks have faucets, drains, pop-up stoppers, tailpieces, and supply lines. Toilets consist of tanks, bowls, fill valves, flush valves, and wax rings. Showers and tubs involve a wide range of valves, heads, spouts, and drains.

Appliances like dishwashers, washing machines, and water heaters also integrate with plumbing. Dishwashers connect via supply lines and drain hoses. Washing machines use hoses and standpipes too. Water heaters have valves, relief lines, heating elements, and thermostats.

Home Plumbing Mastery

Knowing the purpose of each fixture and appliance component helps diagnose issues. For example, if a toilet won't flush, focus on the tank parts responsible for flushing action. If a sink drips, examine the faucet valves and washers first.

Regular maintenance keeps fixtures and appliances working properly. Changing sink aerators and showerheads improves flow. Replacing toilet flappers prevents leaks. Cleaning machine filters and traps prevents buildup. Little fixes prevent big problems. Cosmetic upgrades like new faucets and shower tiles freshen up fixtures too, but don't neglect the critical inner workings behind walls when remodeling. Focus on durability and function, not just form. Learning the optimal use of fixtures also improves plumbing life and reduces waste. Don't run water excessively when brushing teeth or washing dishes. Turn off taps firmly to prevent dripping. Only flush when needed. Use appliances fully loaded.

In summary, fixtures and appliances represent the frontline interaction with plumbing systems. Keep them in good working order through purposeful maintenance and upgrades to enhance daily life. Learn their quirks so you can prevent problems before they occur.

The variety of pipe sizes and materials used in plumbing can seem dizzying to the DIYer. But decoding the terminology and purpose behind pipe types makes your system less mysterious. Let's demystify the meanings behind names like PEX, PVC, ABS, and cast iron. First, pipe sizes. Lengths are measured in feet, but diameters use peculiar units. Copper pipe is sized by trade size rather than inside diameter. So "half-inch" copper pipe has a 5/8" outer diameter and 1/2" inner bore. Plastic pipes use true inner diameters.

Pipe materials each serve specialized roles. Copper is the gold standard for supply lines, praised for durability and water purity. PEX is also common, using cross-linked polyethylene plastic tubing that's more flexible. Galvanized steel pipe is another option but corrodes over time.

For drainage, PVC is the plastic pipe of choice, offering chemical resistance and easy DIY assembly. Cheaper ABS plastic also sees some drainage use but is more prone to cracking. Cast iron remains popular for durability despite heaviness. Drain line sizes don't relate to inner diameter. Instead, width descriptions like 3 x 2 or 4 x 3 refer to dimensions in inches of the oval-shaped cast iron drain cross-section. Confusing!

Vent line sizing gets odder still. Pipe drain table standards assign vent line diameters based on overall drainage fixture units, not actual pipe diameter. It's an archaic custom that persists today.

In summary, pipes use a hodgepodge of size descriptions based on old tradition, not consistency. But decoding the terminology isn't impossible. Memorize the irregular sizing of your home's pipe types through labeling. Then size calculations and replacements become easier. Those baffling pipe labels will finally make sense when repairs come around!

When plumbing emergencies strike, knowing where to quickly shut off water supply and drainage can prevent catastrophic damage. Take time now to locate and label your system's crucial shut-off valves so you're prepared. The main water shut off valve is the first critical control point. It's usually located where the main water service line enters the home, often in the basement, crawlspace, or a utility closet. This valve controls all water flow into the house.

Hot and cold water lines have individual shut-offs as well. The hot water line valve is on the pipe exiting the water heater. For cold water, look for a valve near where the main line branches off into separate cold runs. Individual fixtures also have smaller valve access. Sinks have dedicated valves under or behind them. Toilets have supply valves beneath tanks. Find and label these so you know which fixtures each control.

For drainage, there may be a main shut-off valve on the home's main sewer line where it exits the house. This allows shutting off all drainage outflow at once if needed. More common is finding individual drain shut-offs near each fixture. Bathroom sink drains often have accessible valves. Tubs and showers too. Kitchen sinks may have them under or behind.

Knowing these locations lets you isolate drainage in an emergency if a drain backup occurs. Shut off specific fixtures as needed to reduce damage while troubleshooting the issue.

Finally, water meters usually have an external valve the utility company can use to shut off supply. Make sure this valve remains accessible and functional too. So take a plumbing scavenger hunt now before issues arise. Identify, label, and test valves so you know how to quickly take control during emergencies. Then you can act fast to minimize any resulting mess or hassle.

Home Plumbing Mastery

When tackling any plumbing repairs, consulting the original blueprints and schematics for your home provides invaluable insight. Learning to read these technical drawings unlocks a wealth of system knowledge. On plumbing blueprints, different line styles denote supply versus waste pipes. Solid lines represent cold water supply, dashed lines are hot water lines. Short dashes indicate drainage lines while long dashes show vent pipes.

Arrows on pipelines identify the direction of water flow. Slope direction is also marked to show which way drain lines angle downhill. Circle 'X's highlight the locations of fixtures and drains.

Technical abbreviations label components. Waste and overflow drains are marked as W.O., while bathtub drains show as T.D. Other fixtures contain abbreviations like L for laundry tray, S for sink, and WC for water closet.

Schematics provide more detailed technical diagrams for specific systems like supply grids. These label pipe sizes, water pressure, and knob assignments for fixtures. Elevation callouts show the height above ground for key plumbing elements.

Learning blueprint symbols helps you quickly locate plumbing features and understand how everything connects. When repairs are needed, you can interpret what gets impacted upstream or downstream from the issue based on the designs.

If blueprints aren't available, make your own schematics. Sketch out the supply grid layout with pipe sizes, listing which fixtures each feed. Diagram the main drain lines and vent pipe routing. Include any known slope degrees or elevations. Good documentation provides immense help diagnosing future problems. You can analyze blueprints to identify affected components before demolition even begins, saving effort and minimizing exploratory destruction of home finishes.

So be sure to consult your home's plumbing plans whenever possible. If permits for work are pulled, the AHJ (authority having jurisdiction) frequently requires seeing the updated blueprints too. Keep your documentation up to date.

Chapter 4

Troubleshooting Common Plumbing Problems

Of all the plumbing issues that can arise, leaks may be the most frustrating. But arming yourself with the right troubleshooting approach makes finding and fixing leaks a manageable process. With some diligence and the proper steps, you can tackle even the most elusive drips. Start any leak diagnosis by thoroughly inspecting all visible plumbing components like supply lines, drain traps, and fixtures. Look for obvious wetness, mineral deposits, or water stains which point to a leak's origin. A helpful trick is running colored dye through pipes to try and visualize the leak point. For seeping water inside walls, a moisture meter and infrared thermometer help pinpoint the damp location before tearing up drywall. Match readings against your system diagrams to identify probable pipe runs. Likewise, a faint sewer gas smell can indicate a drain leak behind walls.

To isolate supply leaks, systematically shut off different branch lines and fixtures while monitoring flow at a downstream faucet. If flow stops when shutting one off, the leak is upstream of that point. Narrow down the issue location through process of elimination.

For drainage leaks, monitor background gurgling noises while individually capping off drains. An increase in gurgling indicates that line vents through the leaky section you just closed. By comparing noise changes, you can hone in on the faulty area.

Patience and a methodical approach are key, along with an understanding of your plumbing layout. Never just assume a leak's location—carefully diagnose using visual clues, temperature and moisture readings, isolation valves, and noise changes. Outsmart that drip! With the source identified, make necessary repairs based on pipe type and location. Call a professional if hidden wall excavation is required, but for accessible joints and fittings, implement a fix using this leak detection strategy.

Often just as frustrating as leaks, weak water pressure can be annoying to address, but armed with the right knowledge, you can identify causes and get your flows back to normal. Start by checking the main water supply valve to ensure it is fully open. Partially closed valves are a common pressure culprit. Open outdoor hose bibs all the way too. Inspect supply lines, shower heads, and faucet aerators for clogs or blockages. Flush pipes and replace any obstructed fixtures. Scale buildup in pipes causes pressure loss over time.

Water pressure is often lower on upper floors as water fights gravity. Ensure diameter of upstairs supply lines is sufficient. Larger pipes maintain pressure over distance. For dramatic pressure loss on one fixture, disconnect its supply lines and inspect for blockages. If other fixtures have good pressure, the issue is isolated.

Test water pressure directly at the main supply line before it branches to household plumbing. This reveals losses in your home's supply grid versus overall utility pressure. Compare the pressure here over time to pinpoint internal versus external issues.

Check water meter readings for unusually high or inconsistent flow when fixtures are off. This can indicate serious hidden leaks sapping supply pressure. With wells, inspect well components like pressure switches and holding tanks. Contact a well specialist for deeper diagnostics. Low water tables affect well output.

Finally, keep the water heater set at 120 degrees F or less. Excessive heat causes backpressure and flow loss. By methodically testing pressure from supply inwards, you can isolate pressure drops and zero in on underlying issues. Don't tolerate weak water pressure—take control with a step-by-step diagnostic approach.

Few household noises are as irritating as banging, clanking pipes. They don't just keep you awake; noisy pipes indicate problems needing attention before real damage occurs. Let's troubleshoot solutions for quieting rackety plumbing. Most pipe noises come from excess water velocity or pressure changes. High pressure that accelerates flow can cause vibration and hammering. As valves close suddenly, trapped water smacks pipe walls. Air chambers designed to absorb pressure may be compromised.

Loose pipes are another culprit. Sections that come partly detached from fittings clank as water shifts them. The solution here is re-securing loosened pipe runs to framing with pipe hangers and clamps. Don't let pipes flap freely. Avoid abrupt valve closure that triggers hammering. Replace standard washer valves with slow-closing anti-hammer types. Water should decelerate gradually rather than stopping suddenly.

Consider pressure-reducing valves if supply pressure exceeds 80 PSI. Lower pressure reduces velocity and turbulence. Likewise, verify pumps don't exceed 20 PSI increase if present. Excessive pumping causes noise. For vibrating pipes, pad them with foam insulation to dampen the racket. Rubber acoustic couplings positioned along pipe lengths absorb vibration before it amplifies in metal. Flexible supply tubes also isolate noise.

Finally, clamping floor joists and wall studs absorbs transmitted pipe clamor instead of amplifying it. Use joist insulation hangers and adhesive-backed felt pads to prevent drum-like sound conduction. With some diligent sound isolation and pressure reduction, you can tame noisy pipes and restore peace and quiet. No need for earplugs just to use the bathroom!

Nothing ruins a morning faster than a stubborn clogged drain. Sluggish sinks or backed-up tubs threaten to flood your day with hassle. Arm yourself with the right techniques and tools to vanquish slow drains for good. Start with a handy drain snake, either manual or powered. Feed the rotating barbed wire into the drain opening while turning the crank handle. Advance and retract the snake repeatedly to loosen trapped gunk. Remove any clumps the snake dislodges.

For bathroom sink drains, disassemble the trap underneath and inspect it for obstructions. Hair and soap gunk accumulate here. Rinse out debris and replace the cleaned trap. Likewise, inspect tub drain strainer baskets. Shower drains can be cleared by removing the drain grate and using a flat zip-it tool to hook hair in the drain body below. A tub drain auger fed through the overflow plate access slides past the horizontal trap to clear clogs. A balloon plunger can help dislodge tub clogs. Laundry basin clogs yield to a drain auger fed down the overflow tube to snake through horizontal piping behind the sink. For basement floor drains, remove the grate and use a hand turn auger, feed hose, or CO_2 blaster to dislodge sediment and debris clogging the trap. Knowing the right approach for each fixture saves time and frustration. Always start with the least invasive unclogging method to avoid unnecessary work or risk of damage.

If snakes don't solve the issue, use a plunger vigorously over the drain opening. Form a tight seal with the plunger and plunge forcefully several times to unsettle the clog. Cover overflow drain holes with a rag first.

Follow up with a baking soda/vinegar fizzy reaction. Pour baking soda down the drain, follow with vinegar, and quickly cover the drain. The chemical reaction helps break up organic blockages. Let it sit 30 minutes then rinse with hot water. Prevention is key. Use drain strainers in sinks and tubs to catch hair and food bits. Dispose of grease and oils properly, and pour a monthly maintenance dose of drain cleaner to keep pipes clear.

For deeper clogs beyond your DIY reach, don't hesitate to call a pro. Powerful drain augers can clear blocked drains in walls or floor pipes a homeowner can't access. Let the professional plumbers handle serious backups. Stay vigilant and don't ignore slow drains. A bit of regular maintenance keeps clogs from escalating into huge problems down the road.

A constantly running toilet wastes gallons of water and drives up your utility bill. But with a few replacement parts, you can easily fix a leaky flapper or outdated flush valve causing the issue. First inspect the flapper or seal at the bottom of the toilet tank. A poor seal here leaks water into the bowl, triggering the tank to continuously refill. Look for warping, cracking, or mineral deposits on old rubber flappers and replace if compromised.

Also check the flush valve seat where the flapper seals to verify it's clean. Scrub away any gunk or buildup so the flapper makes a tight seal. A leaky seal here is a common culprit of phantom flushing. For new high-efficiency toilets, a faulty canister-style flush valve could be the issue rather than the flapper. Carefully inspect valve gaskets and O-rings for damage. Replace any worn out rubber components.

If the flapper and flush valve check out, the tank fill valve may be letting water seep into the overflow tube, wasting water down the drain. Adjust the float until water shuts off completely without dripping. Inspect tank bolts holding the toilet down - leaks here mean water is escaping around the base. Tighten bolts carefully to eliminate leakage without cracking the porcelain base.

Finally, replace old 3-5 gallon flush valves with new low-flow valves rated 1.6 gallons per flush. This alone recoups the replacement cost through water savings over time. A bit of regular toilet tune-up and replacement of worn gaskets and washers will keep operational issues at bay before they waste money and water.

Noticing the unpleasant stench of sewer gas in your home? While harmless at low levels, persistent rotten egg smells indicate a problem needing diagnosis. Let's explore fixes for smelly drain vents. The most common source is evaporated water in drain P-traps letting sewer gases up through fixtures. Simply pouring water down unused drains restores the protective water seals.

Incomplete venting can also cause siphoning of traps as air struggles to enter the system. Ensure vents extend above the roof and are clear of blockages. Check for cracked or disconnected vent pipes that allow sewer gas to leak before reaching the roof. Vent pipes should join above drain exits.

A dried out drain trap under washing machines should be filled with water and mineral oil to restore the trap seal and limit evaporation. For septic systems, sewer odors may indicate a full tank needing pumping or a problem with the leach field causing backups. Consult a septic professional. In extreme cases, smoke testing can isolate cracks or flaws allowing sewer gas intrusion. But simple trap refilling solves most sewer smell issues. With some diligent inspection and a few easy repairs, you can banish unpleasant sewer stenches for good. Consult a plumber if DIY efforts don't yield an odor-free home.

Few breakdowns disrupt a household as abruptly as a faulty water heater. But armed with troubleshooting skills, you can get to the bottom of many basic heater issues yourself. Let's walk through diagnosing common water heater headaches. No hot water means checking power first. Verify the breaker or fuse is on and the unit is plugged in, if electric. On gas heaters, confirm the pilot is lit and there's gas flow. Cold supply and hot water valves should be open too.

Insufficient hot water points to heating elements not firing at full capacity. Sediment buildup on electric heating rods inhibits heat transfer. Draining the tank helps clear mineral deposits. Gas heaters may need burners cleaned of debris. Drips usually stem from faulty inlet/outlet valves or corrosion pinholes in the tank body itself. Replace leaky valves. For small tank leaks, drain the unit and attempt a patch using water heater epoxy. If patching fails, replace the whole tank.

Knocking or rumbling noises can indicate sediment accumulation interfering with heating elements. Try draining the tank completely to clear heavy mineral deposits. If noise persists, element replacements may be needed.

If hot water scalds or lacks temp consistency, the thermostat likely needs adjusting. Lower thermostat settings incrementally until hot water temp stabilizes around 120 degrees Fahrenheit.

Finally, if your heater is over 7-10 years old, replacement may be the ultimate solution for chronic issues. Newer heaters are much more energy efficient. When problems persist, it may be time to retire the old tank. With persistence and some DIY maintenance, you can troubleshoot many basic water heater headaches yourself. But for complex electrical issues or tank leaks, call a qualified plumber to properly diagnose and repair the problem.

As we reach the end of this chapter focused on resolving frequent plumbing dilemmas, we hope you now feel empowered to tackle all kinds of leaks, clogs, and malfunctions. While plumbing issues can seem daunting at first, methodically employing the troubleshooting tips provided puts the odds in your favor of diagnosing and fixing many basic problems yourself.

Approach any plumbing misbehavior sensitively and analytically like a detective. Gather clues by replicating the issue and observing closely. Trace back through systems logically to pinpoint the source. Likely culprits are things actively moving or becoming worn like faucet washers, fill valves, flush mechanisms or drain stoppers. But also consider indirect causes like vent blockages, low pressure, and outdated pipes.

Don't rush to solutions before thoroughly understanding root causes. Misdiagnoses waste time and money. The right fix depends on correctly identifying the offending component or blockage location first. Have patience deciphering vague symptoms to get at the origin. Map out the anatomy of your plumbing to understand how flows interconnect.

Home Plumbing Mastery

While repairing malfunctions may seem intimidating as a beginner, have confidence gaining competence with each completed fix under your belt. Invest in a few specialty plumbing tools you'll use repeatedly like valve wrenches, drain augers and supply line flexes. Study tutorials to learn step-by-step repair procedures before diving in. Know when to shut off main water lines to depressurize systems and minimize flooding from mistakes. Call for professional assistance if completely unsure how to approach a perplexing problem.

Remember even seasoned plumbers still encounter mysteries requiring research and trial-and-error. Diagnostics develop through accumulated practical experience. Add tools to your troubleshooting toolkit like motion-activated cameras in drains or pipes to see blockages directly. Use pressure gauges to test pressure differences across plumbing sections. Don't hesitate to call in an extra set of more experienced eyes when stumped.

While some plumbing headaches cause frustration in the moment, view them as opportunities to expand your knowledge. Make repairs teachable moments to gain wisdom for next time. Understanding the anatomy of previous clogs makes future problems faster to clear. Take photos of repairs and keep notes detailing fixes for reference. Invest time upfront to do things correctly so you aren't repeating repairs frequently.

Approach troubleshooting with curiosity of a doctor solving a medical mystery. Meticulousness and creative thinking overcome many plumbing gremlins. And remember all homes have quirky intricacies - don't be intimidated asking questions. Other homeowners and seasoned plumbers have seen it all! With each challenge solved, you gain new skills and confidence to tackle whatever plumbing curveballs arise.

In closing, don't let plumbing issues overwhelm you even when problems feel urgent and pressure mounts. Breathe deep and tackle each situation calmly but deliberately. Avoid hasty guesses that waste resources. Precision in identifying culprits prevents repeated repairs. Your methodical efforts will be rewarded in gaining satisfying mastery of your home's intricate plumbing systems. Be the plumbing detective!

Chapter 5

DIY Faucet and Drain Repairs

Nothing is more annoying than a leaky faucet. Luckily, replacing worn rubber washers and O-rings is an easy DIY fix to get your faucet drip-free again. With a few tools and spare parts, you can stop the seepage quickly. Start by shutting off water supply lines under the sink. Then disassemble the faucet handle and body to expose the internal cartridge or valve parts. Pay attention to component order for reassembly. Examine washers and O-rings for wear, cracks, or compression grooves. These rubber pieces naturally deteriorate over time from the stresses of hot, high-pressure water. Replace any suspect seals.

For compression faucets, unscrew the retaining nut holding the cartridge or valve stem in place. Slide off the worn compression washer and replace it with a matching new one. On cartridge or ceramic disc faucets, pry out the cylinder valve cartridge to access seals and O-rings underneath. Carefully replace seals, lubricate parts, and reinsert cartridge. For ball-type faucets, take out the swivel ball mechanism to swap its worn triangle-shaped inlet seals and O-rings for fresh ones. Make sure to align new seals properly before resetting the ball. Reassemble faucet, turn on water, and test for leaks. Tighten or realign parts if any residual dripping occurs. Add additional plumber's putty or tape if necessary to seal threads.

With this basic seal replacement process mastered, you can save the cost of a plumber visit for routine washer and O-ring repairs. So don't tolerate drips—fix worn seals promptly to prevent bigger problems.

Over time, internal faucet cartridges and ball assemblies wear out from use and need replacement. Don't let stiff handles and leakage ruin your fixture—with a few tools and tips, replacing cartridges and ball assemblies is a straightforward DIY job.

Start by turning off water supply lines and removing the faucet handle to access the interior cartridge or ball parts. Consult your manufacturer instructions for model-specific disassembly. Removing the faucet spout may help create clearance.

For cartridge faucets, pry out the cylinder cartridge using pliers or a specialized cartridge removal tool. Note the orientation for reinstallation. Clean any debris from the faucet body and insert the replacement cartridge.

Ball faucets require extracting the swivel ball assembly using a ball removal wrench or adjustable pliers. Take care not to damage the ball when removing. Lightly lubricate replacement ball O-rings with faucet grease before inserting the new ball. Ensure replacement cartridges or ball assemblies match your model's original parts. Reinstall any washers, seals, or retention nuts to fully seat the new internal components in place. After reassembly, turn on the water supply and test for leaks. Tighten any connections further if needed to eliminate dripping. Reattach spout and handle once operation is verified. Add insulation around pipes to prevent freezing.

With a few specialized tools and basic skills, replacing worn faucet innards yourself avoids the hassle and cost of a full faucet replacement. Save money and get better performance from tired faucets.

Few plumbing problems are as maddening as the drip-drip-drip of a leaky faucet. But with a bit of diagnosis and basic repairs, you can stop faucet leaks in their tracks. Start by determining the leak point. Drips at the spout indicate worn washers or O-rings in need of replacement. Leaks around the handle or base mean loose connections that need tightening. Shut off water supply lines before disassembling the faucet. Protecting the finish, carefully disassemble the handle and body components to access internal parts. Inspect for cracks, corrosion, and wear.

Replace any worn washers or O-rings and lubricate parts with faucet grease during reassembly. These common seals deteriorate over time and need periodic replacement. Match new seals to the originals. For leaks at connections, tighten nuts and screws securing supply lines, spouts, and handles. Add plumber's putty or thread tape to tighten loose threaded joints. Don't over-tighten fragile components. For ceramic disk or cartridge faucets, debris on the valve seat can cause dripping. Scrub the valve seat with an old toothbrush to remove any gunk or mineral buildup.

Finally, replace leaky outdoor hose bib valves. The constant high pressure stresses these valves over time. A new frost-free sillcock faucet repairs seasonal valve leakage.

With a systematic approach, you can tackle most leaky faucet issues yourself. But for stubborn leaks involving complex valve repairs, don't hesitate to call a plumber. A bit of maintenance keeps faucets working great.

Installing a new faucet or tap can update the look of your kitchen or bathroom. With the right skills, you can swap out old faucets for new ones yourself. Let's go over the process. First turn off hot and cold water supply valves and open the faucet to release pressure. Disconnect and detach supply lines and mounting hardware to remove the old faucet. Clean sink surface thoroughly.

Before installing the new faucet, check that the number of sink mounting holes matches the faucet base. Make sure your new faucet includes all gaskets, washers, and fittings needed for reattachment.

Apply plumber's putty around the base of the new faucet to seal the connection. Insert faucet tailpieces through mounting holes in the sink. Place any washers and gaskets then reattach mounting hardware beneath. Connect water supply lines to the new faucet, ensuring a tight seal with pipe tape and wrenches. Make sure hot and cold lines match properly on both faucet and supply valves. The hot line is typically on the left.

Turn on water supply and test for leaks. Tighten any leaky connections. Verify proper hot and cold water flow and pressure through the new faucet. Check draining and swivel motion.

While not overly complex, installing a new sink faucet requires proper planning and patience. Having the right fittings and tools on hand prevents frustration. Take your time and get it right the first time.

Clogged drains are one of the most disruptive plumbing issues. Bathroom sink and tub drains get frequently blocked by hair, gunk, and buildup. Let's review DIY methods for clearing clogs and keeping drains free-flowing. For bathroom sink drains, disassemble the curved P-trap underneath. Place a bucket beneath to catch water. Loosen slip nuts on trap bend sections and disconnect trap arms to access the interior.

Inspect trap components for obstructions. Hair and soap scum often accumulate here. Use needle-nose pliers to extract debris. Scrub interior trap surfaces with a drain brush. Reassemble the freshly cleaned trap. Try a zip-it plastic drain snake for sinks. Feed the barbed plastic strip down the drain until resistance is felt. Retrieve and remove any hair or debris hooked on the strip's spikes. Repeat as needed.

A flexible sink auger fed through the drain opening works deeper clogs loose with its rotating spring tip. Spinning the auger while slowly inserting and retrieving breaks up blockages.

For tub drains, a drain zip-it can hook hair clogs in the crosspiece just below the drain strainer. Remove strainer and extract any loosened debris. Follow up with a tub auger for deeper clogs.

Avoid chemical drain cleaners when mechanical means suffice. But as a last resort, carefully follow directions to clear stubborn organic clogs with caustic drain cleaner products.

With routine maintenance like hair strainer cleaning and monthly drain cleaning product use, you can prevent minor clogs from becoming major headaches.

Pop-up stoppers in bathroom sink drains use a simple lift-and-drop mechanism to hold or release water. But over time, the spring and rubber stopper assembly wears out and needs replacing. Here's how to swap in a new pop-up drain stopper.Start by removing the current stopper assembly. Turn off water supply valves before disconnecting any parts. Unscrew the horizontal rod connecting to the tailpiece. Remove the pivot rod going through the sink drain hole.

Pry off the broken stopper assembly with pliers. Clean excess putty or grime around the drain hole. Assess if the existing drain flange is damaged. Replace flange if needed.

Attach new stopper seal to the drain flange, using plumber's putty to seal underside. Insert new pivot rod through hole in stopper and down through drain. Connect to new horizontal connecting rod. Adjust length of horizontal rod so stopper drops fully when pressed down but rises back up when released. Tighten pivot rod nut and connecting rod to hold adjustment. Test operation several times. If stopper doesn't fully drop open when pressed, the spring tension may need adjusting. Consult instructions to change spring length as needed to achieve full range of motion.

Follow any specific manufacturer instructions for attaching spring, seals, washers or nuts. Proper assembly order prevents leaks or malfunctions. Don't be afraid to call support for guidance.

With a properly installed new pop-up assembly, you can enjoy years of reliable sink drain control. No more stuck plugs or stubborn manual stoppers to wrestle with. While faucets and drains may seem mundane, properly inspecting and maintaining them prevents much larger headaches from water damage resulting from unchecked deterioration.

Remember that methodically tracing issues back to root causes before acting is key, whether dealing with a dripping faucet or sluggishly draining sink. Don't simply replace cartridges or snake drains without diagnosing why problems are occurring in the first place. Look for not only direct issues like cracked washers or clogged pipes, but also indirect contributing factors like water pressure changes, eroded valve seats, or venting problems.

In each repair, go beyond resolving the surface symptom to understand the "why" of plumbing breakdowns. This knowledge equips you to prevent recurrences through proactive steps like adjusting water pressure, replacing aging pipes and fittings, or increasing venting capacity. Preventive maintenance prevents more serious damage and repairs down the road.

Approach all projects with patience and proper preparatory steps. Ensure water is shut off and electrical power is disconnected before any work involving disassembly. Have the right tools for each task readily available. Follow instructions carefully and don't force parts that seem stuck. Call for assistance if completely unsure how to safely proceed. Rushing increases the probability of accidents or botched repairs eventually needing professional re-fixes.

Build competence progressively by taking on more ambitious faucet upgrades like widespread installations or conversion to hands-free models over time. Know your limites—extensive drain re-piping or main sewer line augering, for instance, often requires professional equipment and know-how. But equip yourself to handle as many minor tasks independently as possible. DIY skills pay ongoing dividends.

Make every repair an opportunity to get hands-on education. Catalogue the specific models you have for future parts lookups. Study which seat types, cartridges and washers fit each faucet. Note down drain piping layouts and locations of cleanouts. Such documentation proves invaluable when issues resurface.

Never hesitate to ask more experienced plumbers for consultations when planning major repairs or upgrades. Most are happy explaining best practices and warning of pitfalls before you dive in. Their insights save grief down the road.

In closing, be proud of your growing competence managing home faucets and drains which most impact daily life. Pat yourself on the back for each small win like replacing a leaky gasket or clearing a clogged sink. Know that becoming a proficient plumbing troubleshooter requires perseverance, but you are on the path. With each successful fix, your knowledge and confidence builds. Soon friends and neighbors will seek your wisdom fixing their own faucet and drain headaches!

As we conclude this chapter focused on home faucet and drain repairs, you now have an enriched understanding of these common bathroom and kitchen fixtures along with the skills to tackle many maintenance issues independently. While faucets and drains may seem mundane, properly inspecting and maintaining them prevents much larger headaches from water damage resulting from unchecked deterioration.

Remember that methodically tracing issues back to root causes before acting is key, whether dealing with a dripping faucet or sluggishly draining sink. Don't simply replace cartridges or snake drains without diagnosing why problems are occurring in the first place. Look for not only direct issues like cracked washers or clogged pipes, but also indirect contributing factors like water pressure changes, eroded valve seats, or venting problems.

Chapter 6

Toilet Troubleshooting and Repair

Toilets are the workhorses of home plumbing. Identifying issues quickly leads to faster repairs. Let's review how to diagnose common toilet troubles. A constantly running toilet usually means a poor flapper seal. Listen for flowing water and conduct a dye test by placing a few drops in the tank. If dye seeps into the bowl, the flapper needs replacing.

A weak flush indicates low tank water volume. Check the water line and make sure the fill valve isn't set too low. A fill line lower than 1" above the overflow tube won't create enough siphon during flush. If water intermittently flows into the bowl, the fill valve may be worn out and needs adjusting or replacing. Similarly, a tank that never seems to fully fill could be due to fill valve issues.

A loose base can cause water to leak onto the floor around the toilet. Tighten the bolts anchoring the toilet if wobbling occurs. Adjust them evenly to avoid cracking the porcelain. If the bowl empties slowly or partly fills with disgusting sewer gases, a partial blockage in the drain line is likely preventing proper venting. Plunge the toilet and use an auger to clear obstructions. A backed up toilet overflowing with gross water is every homeowner's nightmare. Turn off the water supply valve immediately to stop additional flooding. Soak up water and call a plumber promptly if DIY snaking can't clear the clog.

By diagnosing problems early based on symptoms, you can quickly zero in on solutions and prevent bigger issues down the road. Stay alert to a toilet's signs of trouble.

The flush valve inside a toilet tank controls the flushing action when the handle is pressed. A failing flush valve leads to leaks and weak flushes, necessitating replacement. Let's review how to swap in a new flush valve yourself.

Start by shutting off the toilet's water supply valve and flushing to empty the tank. Sponge up any remaining water. Unclip and remove the tank lid and set it aside safely.

Disconnect and detach the water supply line from the old flush valve. Unscrew the large locknut anchoring the valve to the tank bottom. Lift out the old valve.

Inspect the opening at the tank bottom where the flush valve seats. Clean away any grime, scale, or debris with a rag. Apply a fresh tank gasket if replacing the old one. Install the new flush valve through the tank opening. Make sure the gasket forms a tight seal on the bottom. Thread on the locknut under the tank and hand tighten. Attach the water supply line to the new valve. Make sure the refill tube entering the overflow pipe is oriented properly. Adjust as needed before fully tightening.

Turn on water supply and let tank fill. Actuate flush handle several times to test operation. Adjust valve per instructions if needed to achieve proper flush volume. With a regular flush valve replacement, you can restore a strong, leak-free flush to your aging toilet. Don't settle for constant leaks and frustrating ghost flushes!

The toilet tank contains components that work in unison to control flushing action. Failures can lead to leaks and weak flushes. Replacing worn tank parts restores proper operation.

The flapper is a key component—its seal with the flush valve prevents water from flowing to the bowl. A leaky, warped flapper allows phantom flushing and needs replacement. Match your original size and style. The tank ball float controls the fill valve flow shutoff once the tank reaches the correct water level. If water overfills the tank, adjust the float lower on its arm or replace it if worn out.

The fill valve delivers water into the tank through a refill tube. These valves can wear out over time, requiring replacement. Match the gallons per minute rating of the old valve.

The lift chain connects the flush lever handle to the flapper. A disconnected or overly slack chain impedes full flapper lifting. Reconnect or replace the chain. The overflow tube prevents water from overfilling the tank. Make sure no obstructions block it. Replace the tube or refill tube if cracked.

Finally, make sure tank bolts are tight. These anchor the tank to the bowl. Loose bolts allow leaking and movement. Tighten or replace deteriorated bolts. With a few replacement parts, you can restore any aging toilet's tank components to proper performance. Don't settle for lackluster flushes!

Over time, toilet seats and lids can crack, warp, or just become unsightly. Luckily, replacing them is a quick DIY project needing only a screwdriver and a few minutes. Start by shutting off the toilet's water supply valve and flushing to empty the tank. Remove the old seat and lid by unscrewing the bolts underneath on each side. Set bolts aside if reusable.

Inspect the bowl rim for any remaining particles, putty, or grime. Clean the rim completely to ensure a proper seal. Wipe any water left in the bowl. Position the new seat and lid on the bowl. Insert new bolts or reuse existing bolts if still functional. Add extra washers if the bolts are too long.

From above, tighten each bolt evenly until snug. Avoid over-tightening to prevent cracking the porcelain. If necessary, cut the end of the bolt with a hacksaw before adding the cap. If the seat and lid don't sit flat or rock after tightening, try shimming under them with plastic or metal washers. This stabilizes the components.

Finally, apply a thin bead of silicone sealant around the underside of the toilet seat. This provides extra protection against leakage and wobble. Let the caulk cure overnight before use.

With proper installation, your fresh toilet seat and lid should serve you well for years before needing another quick replacement. Enjoy the renewed style and comfort!

The wax ring underneath the toilet creates a water-tight seal between the toilet and drainpipe. A damaged or leaky wax ring allows smelly gases and waste to escape. Replacing this critical gasket yourself maintains a sanitary barrier. Start by shutting off the toilet's water supply and flushing to drain the tank completely. Sponge up any remaining water in the bowl. Remove the toilet tank lid and set it aside. Unbolt the tank from the bowl by loosening the nuts on each side. Carefully set aside the tank.

Unbolt the toilet bowl from the floor. Situate it so you can access the wax ring underneath. Scrape off the old wax ring down to bare flange. Clean the area thoroughly. Inspect the closet bolts, flange, and floor horn for any damage. Make repairs before installing the new wax ring. Lubricate the horn with a thin layer of plumber's putty.

Position the new wax ring evenly over the horn, with the taper pointing down. Apply even pressure to fully adhere the ring and seal its bottom edge. Carefully lower the toilet bowl in place, with the horn centered in the ring. Rock side to side slightly to imprint the ring. Reinstall bolts and attach tank.

Turn on water supply and test operation. Check for leaks under toilet base. Give the wax ring adequate time to fully adhere and seal before using. With regular inspection and replacement, the humble toilet wax ring maintains a critical odor and germ barrier, letting you enjoy a sanitary, leak-free throne.

Bidets and toilet sprayers add a refreshing cleansing element to the bathroom. But like any plumbing fixture, they can develop issues over time. Let's review troubleshooting common bidet and sprayer problems.

A bidet not spraying water likely indicates a clog or problem with the supply line. Check shut-off valves to ensure water is reaching the bidet. Clear any debris blocking nozzle holes. Low spray pressure might mean sediment buildup in supply hardware or nozzles. Disassemble and clean parts of any flow-blocking debris. Adjust faucet cartridge or replace if worn. Leaks between the bidet and toilet bowl suggest a damaged gasket. Verify mounting bolts are snug. Remove bidet and replace gasket with a new self-adhesive foam or rubber gasket.

If water sprays erratically from the bidet, the nozzle direction may need adjusting. Consult manual for realigning nozzle angles to proper direction.

For attached sprayers, leaks near the mounting nut indicate worn O-rings. Disassemble sprayer from hose and replace O-rings to stop dripping. Verify tight nut mounting. If sprayer buttons stick and don't spring back, remove handle and clean out sediment or buildup behind button. Lubricate parts before reassembling.

Odd odors might mean mold or mildew buildup. Disinfect and clean all surfaces and nozzles with bleach spray and scrub brush. Flush water lines. By isolating issues and components, you can restore your bidet or sprayer to full soothing function. Don't live with unpleasant sanitation devices!

Chapter 7

Hot Water Heater Repairs:

Understanding what's happening inside your water heater demystifies repairs and maintenance. Let's overview the key components and heating process that deliver hot water on demand. Cold water enters the tank through an inlet pipe into the bottom. Inside, electric heating elements or gas burners heat the water. As it warms, the water rises through convection towards the top outlet. A thermostat and thermometer regulate temperature. The thermostat adjusts heating elements or gas flow to maintain the set temperature. Insulation minimizes heat loss from the tank exterior.

At the bottom, sediments settle out over time. An anode rod attracts corrosion instead of the tank. A drain valve allows flushing out buildup through routine maintenance.

Relief valves provide safety release points if excessive pressure builds up. One valve connects to a discharge pipe directing hot water safely away. The other allows air intake. Hot water exits the top outlet when a fixture is opened. As hot water leaves, incoming cold water displaces it, keeping a constant supply hot and circulating.

Knowing the heating process and components helps inform repairs. For example, insufficient hot water may be due to a faulty thermostat or clogged heat elements. A systematic approach prevents misdiagnosis. So take time to understand what's happening inside your white-boxed water heater. A little learning goes a long way to maintaining hot water convenience.

To keep hot water flowing safely, water heaters need some periodic maintenance. Let's review easy DIY tasks like sediment flushing, anode rod inspection, and pressure relief valve testing.

Flushing the tank removes mineral sediment that impedes heat transfer and shortens lifespan. Simply drain the tank via the spigot, then refill it to wash loose deposits out. Flush annually or per your manual. Check the sacrificial anode rod, usually every 2-3 years. This rod attracts corrosion instead of the tank. Unscrew it to inspect wear. Replace if eroded over 50%. Use pipe tape on threads when reinstalling.

Test the TPR (temperature and pressure relief) valve annually by briefly lifting or pressing the test lever. This ensures water can discharge if overheating or overpressure occurs. If it fails to open, replace the valve. Examine pipes and fittings for leaks or corrosion. Tighten loose joints or replace corroded pipes. Check that TPR discharge pipe allows full drainage without clogs.

Inspect electric elements for calcium buildup or cracks indicating failure. Descale rods and replace if damaged. Check thermostats and wiring connections too.

For gas heaters, examine the draft hood and thermocouple. Clean any spider webs or debris from the hood. Replace corroded thermocouples. With this simple routine maintenance, your water heater will operate safely for years past its warranty expiration. Don't take hot water for granted!

Electric water heater repair often focuses on the heating system—the thermostats and elements that heat the water. Replacing these parts restores efficiency and hot water capacity. The thermostat governs temperature by cutting power to elements when the set temp is reached. If temperature fluctuates widely, the thermostat likely needs replacing. Turn off power before removing it. Match the new thermostat's temperature range and power capacity to the original. Some are direct screw-in replacements, while others require rewiring terminals. Consult diagrams.

Electric heating elements consist of metal rods or spiral coils that heat via resistance when powered. Sediment coats elements over time, hindering heat transfer. To replace elements, first drain the tank and shut off electrical supply. Remove access covers and detach wire connections. Unscrew the element from the tank.

Avoid scratching the tank surface when coaxing out old elements. Use penetrating oil if needed to loosen the threads. Apply pipe joint compound to new element threads before screwing it in.

Double check that replacement elements match original wattage specs. Reattach wires securely with corrosion-proof connectors. Refill the tank once elements are installed.

With some electrical savvy and basic handy tools, you can give aging water heaters new life by replacing worn thermostats and elements yourself. Enjoy hot water for years to come!

Few plumbing issues are as alarming as a leaking water heater. But many leaks can be fixed with basic tools if you act quickly. Let's review common electric and gas water heater leak repairs. Leak sources include tank seams, pipe connections, fittings, relief valves, and tank components. Inspect all areas carefully to detect the origin. Catching it early makes repairs simpler. For small seepage at tank seams, drain the tank and attempt a patch using water heater epoxy. Thoroughly clean and dry the area first. Apply epoxy per kit instructions.

If leaks originate at pipe threads, try re-sealing with thread tape or pipe dope compound. Tighten fittings moderately. Excessive force can crack plastic fittings. Corroded or stripped fittings should be replaced, not reused. Use pipe wrenches to detach damaged fittings. New fittings must match the pipe size and material.

Weeping at relief valves likely means a bad seat seal. Replace relief valves that display any water leakage. Follow exact specifications for temperature and pressure rating.

For major tank leaks not patchable, replacement is needed. An old tank over 7-10 years has often reached the end of its service life anyway. Know your main water and power shut offs so you can act quickly in an emergency. But even slow leaks can often be repaired with patience and the right materials. Don't panic!

To extend your hot water heater's life, periodic draining and flushing keeps sediment from accumulating. Let's go over the steps for a proper tank flush. Start by turning off power to an electric heater or the gas line for a gas heater. Then shut off the water supply valve. Attach a garden hose to the tank's drain spigot.

Open the pressure relief valve at the top of the tank. This allows air in so water can drain out the spigot freely. Place the hose's end in a bucket or drain area. Open the drain spigot using channel-lock pliers or a crescent wrench. Water will rush out heavily at first. Let it completely drain until just a trickle remains.

Once drained, close the spigot and remove the garden hose. Turn the water supply back on to refill the tank. The influx of clean water will flush loose sediment out through the open drain valve. When water runs clear from the drain valve, close it. Allow the tank to fill completely, then restore power or gas. Bleed air from pipes by opening fixtures until water flows steady.

Aim to flush annually or per manufacturer guidelines to maximize efficiency and tank life. Schedule it when hot water needs are minimal. Proper flushing removes scale-causing minerals.

With this simple maintenance completed, your hot water heater will benefit from less corrosion and optimal heat transfer. Take preventative care of your water heating workhorse!

When is it time to retire your aging, problematic water heater and buy a new one? Here are signs indicating replacement rather than repair may be the wise choice. If your heater is over 10 years old, replacement may make more sense than repairs. New models are much more energy efficient, saving energy costs long-term. Continuing problems with sediment buildup, strange noises, and corrosion despite repairs point to a failing tank. At some point, patching leaks and replacing parts becomes wasteful.

Inability to keep water at a stable, comfortable temperature indicates thermostat and heating element failures beyond DIY repair. Time for a modern upgrade. If simple maintenance steps like draining and flushing no longer resolve issues, serious internal damage likely exists. Trying to salvage an old tank may just delay the inevitable. For major leaks, especially around tank seams, comprehensive repairs involving tank welding or heavy patching exceed the cost of replacement. Don't throw good money after bad. If new building or energy codes in your area make your older heater non-compliant, replacement may be mandatory. Get a head start and upgrade voluntarily.

Compare the cost of your latest repair bill against the price of a new heater. At a certain point, replacement provides more value despite the upfront cost. If you've gotten a decade of faithful service from your water heater, there's no shame in retiring it for a newer, more efficient model. Know when to say goodbye!

You now have an enriched understanding of gas, electric, tankless, solar and specialty models. Properly installing, inspecting, troubleshooting, servicing and replacing water heaters prevents untimely failures and expensive water damage.

Remember that preventative maintenance is crucial for a long functioning lifespan. Drain and flush heaters annually to prevent sediment buildup which can cause serious issues like combustion chamber cracks or element failures. Test pressure relief valves and anodes rods during flushing and replace if needed. Also inspect vents and exhaust piping for blockages or leaks that impact efficiency.

When operational problems arise, methodically trace symptoms back to root causes before acting. Consider both direct issues like failed thermostats or elements, but also indirect factors like faulty vents, incorrect gas pressure, or high mineral content in supply water. Take a holistic view of the heating system to prevent recurrences.

Ensure electricity and fuel sources are disconnected before any disassembly or component testing during diagnosis. Rushing into repairs without isolating systems risks electrocution or fires. Patience also prevents misdiagnoses and wasted effort. Have specialized tools ready for tasks like element removal or draining. Follow instructions carefully and call for assistance if unsure how to safely proceed.

Build competence progressively by taking on more ambitious upgrades like tankless water heater installation over time. Familiarize yourself with differences in venting, electrical load and flow dynamics compared to traditional models. But know your limits—extensive gas line alterations or HVAC integration often requires professional expertise and permitting. Focus DIY skills on maintenance and minor repairs.

Make each fix an opportunity to get hands-on education. Catalogue the specific models you have for future parts lookups. Note down piping layouts, fuel sources and electrical connections that influence serviceability. Such documentation proves invaluable for preventative maintenance and when issues resurface.

Never hesitate to ask experienced plumbers or HVAC technicians for consultations when planning major heater upgrades or repairs. Their insights help avoid pitfalls and code violations. Be willing to pay fair pricing for quality work sustaining your home's hot water nervous system.

Home Plumbing Mastery

In closing, take pride in your growing competence maintaining a critical utility many take for granted. The skills you've developed solve comfort and sanitation issues quickly, prevent property damage, and may even save lives by averting carbon monoxide accumulation from faulty heaters. Reward yourself after each small win repairing a thermocouple or replacing heating elements. Soon family and friends will seek your wisdom ensuring their endless hot showers!

Chapter 8

Garbage Disposal Repair

Garbage disposals are invaluable kitchen appliances, but how do they grind food waste so effectively? Understanding the internal parts and mechanisms makes repairs straightforward.

Food scraps enter the disposal through the sink drain opening into a grinding chamber. At the bottom of this chamber, a rotating metal turntable plate connects to the grinding ring via protruding lugs.

As the turntable spins at high speed, lugs drive the grinding ring around the perimeter of the chamber in the opposite direction. This action creates a shearing force that pulverizes food particles against the stationary grind ring wall. The shredded bits are flushed down through openings along the perimeter into the discharge chute. Water from running faucets aids this flushing process. Spinning impellers help push waste down while blocking backflow.

The spinning turntable and grinding ring are powered by a motor mounted on top of the unit. A wall switch activates the motor's high torque to start the grinding process when needed.

Some high-end disposers have special features like quieter operation, corrosion resistance, or dual grind stages. But the basic grinding action remains the same across most common residential models. Knowing how disposals operate informs repairs. A jammed turntable, damaged grind ring, worn lugs, or leaky chamber will undermine disposal operation. Fixing these issues keeps food waste disappearing efficiently.

Clunks, hums, and jams—garbage disposals can develop a few common mechanical issues. But don't despair—most problems are fixable with basic troubleshooting and maintenance. A disposal that makes loud grinding noises could have a broken motor mount or loose components. Verify mounting bolts are tight. Try tightening the flywheel via the bottom access if noises persist while running.

Leaks between the disposal and sink likely indicate a damaged sink flange seal. Shut off power and verify the flange is fully tightened. Replace the flange seal with a compatible plumber's putty or rubber gasket. If the disposal hums but won't grind, the flywheel is likely jammed. Use an Allen wrench in the bottom access port to manually turn the flywheel back and forth until it spins freely again.

A tripped or clicking reset button often means something is obstructing the flywheel impellers. Unplug power, then carefully remove obstructions through the disposal drain opening. Avoid reaching hands inside. If water doesn't drain well, check for clogs in the discharge chute, sink drain piping, or trap. Clear jams with a drain snake or by unscrewing and cleaning components.

Foul odors are typically caused by food buildup. Deodorize with baking soda and lemon and avoid putting overly starchy or fibrous foods down the disposal. With some informed troubleshooting and DIY maintenance, most disposal issues can be quickly diagnosed and repaired. Just take precautions and never stick hands blindly into a running disposal!

Is your disposal humming but not operating? Chances are the internal flywheel is stuck. With the right technique, you can unjam a seized disposal quickly. First unplug the disposal unit for safety. Confirm the reset button is not popped out. If it is, push it back in to reset the motor circuit. Insert an Allen wrench into the bottom inspection port. Turn it back and forth to manually rotate the flywheel and release any stuck grinding ring or turntable.

For stubborn jams, insert a wooden broom handle into the top disposal opening. As you rotate it, the handle may forcefully dislodge trapped materials freeing up the flywheel. Another trick is to repeatedly insert and remove a sturdy wooden spoon through the disposal drain opening while pressing the reset button. This may jostle loose any debris lodged in the turntable.

Avoid metal objects like forks or wire hangers to unjam a disposal. This can damage the internal components. Use only wood objects to poke safely inside. If water stands in the disposal, use a turkey baster to suck it out. Accumulated water adds weight, making it harder for jammed mechanisms to turn. Confirm adequate power supply to the unit. Low voltage can cause slow operation and jams. If needed, inspect wiring for loose connections. With some improvised tools and persistent maneuvers, most flywheel jams can be freed without a plumber. Just never stick fingers into an active disposal in attempts to unclog!

Garbage disposals contain several components that wear out over time and require replacement. With some disassembly skills, you can swap out common parts like motors, grind rings, and splash guards. A grinding noise when activated indicates potential issues with the motor, flywheel, or shredder ring. Unplug power and remove the lower mounting ring to inspect. Replace damaged parts matching original specs.

Leaks near the top perimeter likely mean the sink flange seal is worn out. Shut off power and drain the trap. Disconnect drainpipes and wiring to remove disposal. Replace the plumber's putty or rubber gasket seal. A damaged or detached splash guard allows debris to fling out while grinding. Unplug the unit and snap a new properly sized splash guard into the outer edge of the disposal opening. If the grinding chamber develops cracks or corrosion holes from exposure to food acids, replacement may be necessary. This involves fully disconnecting all piping and wiring from the unit.

Slow grinding can also point to a worn shredder ring or blunt turntable lugs. Check these components through the bottom inspection port. If deteriorated, detach and replace. Reinstall all components properly before restoring power. Make sure grind rings, gaskets, and guards are fully seated and secure. Take your time to avoid leaks. With some handy disassembly skills and replacement parts, you can often renew a dysfunctional disposal yourself instead of purchasing a new one. Just take precautions when working in the cabinet space beneath.

When an old disposal finally dies, installing a new one is a fairly straightforward DIY task. With some electrical know-how and the right materials, you can replace your unit for a fraction of the cost of hiring a plumber. Start by unplugging and disconnecting the old disposal's drain pipe, mounting assembly, and electric wiring. You may need to detach the sink strainer and P-trap for easier access. Prep the new disposal by threading a wire connector onto the power cord and screwing on the discharge tube and mounting flange. Apply plumber's putty beneath the flange.

Insert the new disposal up through the sink drain opening while simultaneously lining up the mounting ring holes with those drilled in the sink. Hand tighten the mounting ring bolts. Connect the dishwasher discharge tube to the disposal inlet (if applicable). Attach the drainpipe to the disposal discharge tube, tightening the slip nut flange. Reinstall P-trap. Use wire nuts and electrical tape to connect the disposal's wire harness to the electrical supply wires in the cabinet. Follow disposal instructions carefully. Once fully assembled, test the installation. Slowly fill sink bowl while running water and flicking on disposal in short pulses. Check for leaks or uncommon noises.

Finally, run cold water and turn on the disposal while gradually adding dish soap. This helps season the new grinding components to reduce corrosion. With some perseverance and attention to safety, installing a new garbage disposal is an achievable DIY weekend project. Just never attempt repairs on an active unit still plugged in.

You now have an enriched understanding of these useful kitchen appliances that pulverize food waste. Properly installing, operating, inspecting and servicing disposers prevents many headaches of clogged drains.

Remember that mindfulness in what goes down the drain determines longevity. Avoid putting fibrous material like peels, eggshells and coffee grounds into disposals which can wrap around the impeller and jam the motor. Run plenty of cold water before, during and after grinding to flush waste fully into drains. Give the motor time to wind down between batches.

When operational problems arise, methodically trace symptoms to root causes before acting. Consider direct issues like stuck impellers or worn grind rings, but also indirect factors like motor overheating from heavy use or voltage problems. Take a systemic view of the appliance and hookups to prevent future breakdowns.

Ensure electricity is disconnected before any disassembly or component testing during diagnosis. Rushing into repairs risks electrocution or cuts from blades. Have the right disposer-specific tools ready for tasks like resetting circuit breakers, spin tests and removing jams. Follow instructions carefully and call for assistance if unsure how to safely proceed.

Build competence progressively by taking on more ambitious upgrades like three-stage disposer installation or conversion to quieter models over time. But know your limits – complex wiring alterations or sewer line access often requires bringing in electricians or plumbers. Focus DIY skills on minor maintenance and easy swaps.

Make each fix an opportunity to get hands-on education. Catalog the specific model you have for future parts lookups. Note down critical dimensions and hookup specifics that will prove invaluable when issues resurface. Maintain a dedicated toolkit just for disposer service.

Never hesitate to ask appliance technicians or plumbers for consultations when planning major disposer upgrades or diagnosing tricky repairs. Their insights help avoid safety mishaps, code violations or improper installations. Be willing to pay fair rates for quality work on these useful appliances.

In closing, take pride in keeping this workhorse appliance functioning smoothly for years through proper maintenance habits and timely repairs. With the skills you've developed, you can now quickly resolve nuisance jams, humming sounds, tripped breakers or diminished grinding. Your efforts prevent backed-up sinks and ensure ongoing convenience of waste disposal. Soon enough family members will seek you out as the household disposer guru!

Chapter 9

Installing and Replacing Plumbing Fixtures

Upgrading or replacing plumbing fixtures like sinks, tubs, and toilets enhances your home. Careful planning makes installations proceed smoothly and safely. Let's review key steps to plan your project. First, assess the existing fixture. Note the dimensions, configuration, water supply and drain connections. This informs the requirements for its replacement. Take measurements and photos for reference.

Determine if you'll need additional repair work like re-tiling, wall repair, or relocating supply shut offs. Factor these into project scope and timing. Minor drywall or paint work may be required.

Pick out the new fixture based on measurements, utility hookups, and design needs. Compare options at home improvement stores. Avoid ordering online unless you can verify dimensions.

Ensure your plumbing can accommodate the new fixture. Double check space, drain placement, water supply pressure, and venting. Consult original building plans if possible.

Assemble all necessary materials beforehand: new fittings, valves, seals, adapters, etc. Nothing slows a project like mid-stream trips to the hardware store. Carefully review manufacturer instructions for the new fixture. Check for special considerations like framing reinforcements or non-standard installations. With upfront planning, you can handle a fixture swap efficiently. Set a schedule allowing ample time. The key is avoiding surprises and delays once the deconstruction begins.

When replacing sinks, tubs, or toilets, the first challenge is safely dismantling the old fixture from plumbing connections. With care and proper techniques, you can remove old fixtures without issue. Always start by shutting off water supply valves and draining any residual water from the fixture. Disconnect and unthread supply lines from valves or tees. Use a bucket to catch any spillage.

Detach the drainpipe from the waste tee or stub out. A pipe wrench or channel locks work for unscrewing slip nut joints. Place a bucket beneath to catch water. Remove any anchor screws, brackets, or adhesives that secure the fixture to the wall or flooring. Scrape away old putty, sealant, or gasket material with a putty knife.

For toilets, empty the tank and disconnect the supply line. Unbolt the tank and bowl from beneath before lifting the toilet. Place shims beneath the bowl if no floor flange remains.

With cast iron tubs, remove the old drain shoe and unthread or cut old drainpipes. Break sealant bonds with careful prying and twisting force. Seek assistance moving very heavy tubs.

Take care not to damage water supply valves or drain/vent lines that will be reused. Photograph connections before disassembly in case of confusion later. Check for hidden supply shut offs or valves once the fixture is fully detached. Turning these off minimizes confusion about any continuing drips or leaks during the transition period. With patience and proper preparation, you can remove old fixtures without issue. Just work methodically and carefully to avoid damaging adjacent plumbing lines.

With some DIY skills, you can install a new sink or vanity to improve your bathroom's form and function. Careful planning ensures your new fixture fits properly and installs cleanly. Gather all necessary materials beforehand: new faucet, strainer, P-trap, supply tubes, shut-off valves, sealant, tools, etc. Ensure the vanity countertop opening aligns with your new sink. Start by shutting off water supply valves and disconnecting supply lines. Remove the old sink and clean the vanity top contact area. Apply a continuous bead of silicone sealant around the sink rim.

Lower the new sink into place while aligning any pre-drilled faucet holes. Position as needed, then press down firmly to adhere the sink into the sealant. Wipe away any excess sealant squeeze-out.

Reconnect supply lines to valves and insert the strainer and P-trap into the sink drain opening. Tighten from below if needed to prevent leaks. Test for leaks.

If swapping vanities, first disconnect plumbing and remove the old vanity. Place the new vanity and re-plumb supply lines and drainpipe. Adhere to wall if needed. Carefully install the new faucet and handles according to manufacturer instructions. Make sure supplies are firmly attached without drips or leaks. Take your time orienting the sink properly on the first try—adjustments after sealing can break the watertight bond. With care, you'll enjoy a stunning new bathroom focal point.

Installing a new bathtub or shower enhances your bathroom's relaxation and functionality. With proper planning and materials, you can handle replacing old tubs yourself. Measure your bathroom space carefully to ensure the new tub or shower enclosure fits the footprint. Address any framing or wall surface changes needed before installing.

For tubs, determine if you want a drop-in or undermount installation. Remove old tub and plumbing, then adapt supply and drain lines as needed to fit the new tub. Extend lines for alcove-style tubs. Set the new tub in place and mark plumbing holes. Drill access holes at marked spots. For undermount tubs, set tub in adhesive on a ledge then seal edges to form a watertight seal.

Attach supply pipes to valves and shower head. Install the tub drain assembly, connecting the stub out to the tub shoe. Seal joints with thread tape and pipe dope compound to prevent leakage. For showers, check if the floor needs reinforcement to support the enclosure weight and water. Install backing materials on studs before attaching the shower pan and walls.

Carefully apply waterproofing around the shower interior. Slope the pan slightly towards the drain and seal all joints. Install the drain assembly. Finish plumbing supply lines and shower head feeds once enclosure framing is complete. Perform water tests to check for leaks before closing up walls. Take extra time preparing the surrounding environment for wet conditions with tiles, wall protectors, and water-resistant drywall. Plan meticulously!

Installing new toilets or bidets improves your bathroom's sanitation and comfort. With proper planning and equipment, you can handle replacing old fixtures with new ones. Double check rough-in measurements for the toilet/bidet before purchasing. Note supply valve and drain locations relative to the flange. Buy compatible fixtures.

Shut off water supply and remove old fixture. Plug drainpipe before working to prevent sewer gas escape. Scrape old wax, bolts, and seals until flange is clean. For bidets, verify mounting studs align with bidet pre-drilled holes. Place the bidet atop the flange orienting the supply and drain connections properly. Anchor with washers and nuts.

Apply wax ring evenly around toilet horn, tapered end down. Gently set bowl atop flange, pushing down lightly to imprint wax. Insert anchors through bowl into flange. Alternatively, use a wax-free gasket seal for the toilet. Ensure bowl outlets align before adhering gasket and anchoring toilet. Attach supply line to valve or bidet T-connector. Run water and check for leaks. Reseal if any drips occur before completing the installation.

Caulk toilet/bidet base after confirming proper operation. Let cure overnight before heavy use. Add fixtures like grab bars as needed. With the right planning and preparations, you can tackle toilet and bidet replacements yourself. Just work carefully to align new fixtures properly without damage.

68

Chapter 10

Advanced Leak Detection and Pipe Repairs

Pinpointing leaks inside walls, floors, and ceilings can be challenging. But with the right equipment and a systematic approach, you can zero in on the location of hidden plumbing leaks. Start by examining areas around appliances and fixtures for signs of dampness or dripping. Stains on walls or the underside of cabinets indicate a leak could be near. Use a moisture meter to detect elevated moisture levels in drywall or paneling. Test along pipes runs, and where higher, begin carefully cutting away coverings to access the leak area.

Infrared thermometers also identify temperature differences caused by water leaks versus surrounding materials. Scan slowly, watching for hot or cold spots. Check for musty sewer gases which suggest water is escaping from drain pipes into surrounding spaces. Follow your nose to determine the source area.

Turn on bathtub and sinks faucets to pressurize drain lines. Listen in walls for the sounds of dripping water. Isolate the area before cutting open. For supply pipes, systematically shut off different lines while checking a downstream faucet. If flow stops when one valve is closed, the leak is upstream from there. Cut exploratory holes where evidence points, starting small. Opening up walls in multiple areas may be needed to finally reveal the exact leak location. Patience and meticulousness are key. Leaks can travel along pipes and beams. But with diligence, you can pinpoint any hidden plumbing leak.

Leaks in supply lines inside walls create major water damage if unaddressed. With some drywall surgery and the right parts, repairing embedded supply line leaks is a manageable DIY project. Carefully open up the wall area around the leak by cutting away drywall using a razor knife. Shut off nearby valves to minimize water spurting while exposed. Thoroughly inspect the damaged pipe area to determine the extent of the leak. Copper and PEX pipes can often be patched, while polybutylene or steel should be replaced.

For small pinhole leaks in copper, sand off mineral deposit buildup around the leak to expose fresh pipe. Apply waterproof epoxy putty, smoothing it over the puncture. Allow proper curing time before retesting. With PEX tubing, use repair couplings that seal around the damaged section of pipe. Cut out the leaky portion and position coupling so it overlaps either end by at least 1 inch.

If pipe replacement is needed, measure the damaged section and cut out using a pipe cutter or hacksaw. Splice in a similar material and length using couplings. Where pipes intersect with joints and elbows, leaks may require replacing entire fittings. Make sure to prime and cement new plastic fittings properly for a tight seal. Insulate repaired segments before patching walls to prevent refreezing. Flush lines thoroughly post-repair to prevent debris clogs downstream.

With some skill and the right materials, you can tackle embedded supply line leaks yourself instead of resorting immediately to re-piping. Just be sure to test repairs thoroughly before closing up walls.

Leaks inside enclosed walls and ceilings cause water damage and mold if left unaddressed. With some demolition work and patience, you can fix embedded leaky pipes before major destruction occurs. Start by shutting off water supply valves to the area and draining lines. Open walls and ceilings along the pipe path to expose the leak location. Turning off power is wise for ceiling leaks.

Thoroughly inspect the damaged pipe area to gauge the repair scope. Look for corrosion pitting, cracks, separated joints, or perforations. Copper and PEX pipe leaks can often be repaired with epoxy putty patches or splice couplings. If the pipe is too deteriorated, replace the damaged section with an identical size and material. For supply pipes behind toilets or cabinets, consider rerouting replacement pipe sections through more accessible areas to simplify future repairs. If drain or vent pipe leaks, detach at joints to replace cracked sections with ABS or PVC plastic pieces cleaned with primer and cemented.

Insulate repaired pipes before enclosing walls. Upgrade insulation throughout the system to prevent freezing. Flush lines to clear loose sediment post-repair. Before patching walls, verify repairs are watertight under system pressure. Lingering leaks could reappear after concealing. Take time to test thoroughly. With diligence and the right materials and couplings, most in-wall pipe leaks can be remedied by DIYers. But know your limits—significant corrosion may warrant re-piping.

Few plumbing emergencies are as serious as a burst pipe. But if you act quickly, turn off water, and prep materials, repairing burst pipes becomes a manageable DIY job. First, shut off the main water supply valve. This should be an obvious large lever or knob near your home's main pipe inlet. If unsure, start turning off all valves until water shuts off.

Drain the plumbing system completely by opening fixtures on the lowest and highest points of your home. With water evacuated, leakage will decrease. Inspect the burst section. Copper and plastic pipes can often be patched, while repairs on galvanized steel or lead should be left to a pro. Assess if replacement is needed.

Thoroughly dry the area with rags and a hairdryer. Remove any loose pipe pieces. Sand damaged ends of copper pipe to expose fresh metal for bonding patches or couplings.

Apply epoxy putty in a thin layer around the damaged section, using the included putty knife. Reshape once fully hard for at least an hour before turning water on.For larger bursts, cut out the broken section with a pipe cutter. Replace with new pipe of similar size and material using couplers. Apply pipe joint compound to metal pipe threads.

Insulate all repaired areas before restoring water flow. Open valves slowly and check for leaks. Further epoxy patches or adjustments may be needed as pressure increases. With preparation and prompt attention, burst pipes don't have to be catastrophic. Know when to call a pro, but many projects are DIY-friendly with the right tools.

For quick emergency pipe repairs, metal clamps and wrapping patches offer temporary solutions until permanent fixes can be made. Let's look at best practices for using pipe bandages. Pipe patches come in coil rolls of adhesive-backed fiberglass or mesh material. Clean damaged pipe thoroughly before applying patches. Remove any loose pipe pieces or debris. Wrap patches around the leaky section with 2-3 layers, pressing firmly to adhere. Extend wrapping 2 inches beyond damage perimeter. Reshaping may help conform patch to pipe curves.

For large holes up to 3 inches wide, use rigid fiberglass or stainless-steel plate patches with gaskets. Bolt them in place over damage using integrated tabs or bands. Water pressure helps seal the rigid cover. Stainless hose clamps can also seal many pipe punctures or cracks. Use screw-tightening clamps to compress leaky areas and stop flow. Isolate damage best as possible between clamps.

Size clamps appropriately for your pipe diameter, tightening them moderately. Very tight clamping can worsen pipe cracks. Use cloth insulation beneath to protect pipe finish. For seepage at joints, apply epoxy putty before wrapping patches around the area for added durability and waterproofing. Patches and clamps are temporary fixes only. Replace damaged sections properly as soon as possible. But in a pinch, these emergency bandages stop leaks fast.

When pipe sections become too corroded or damaged to repair, replacement is necessary. With some demolition work and new materials, you can often replace bad pipes yourself. Determine the pipe material, diameter, and length needing replacement. Measure carefully to get an identical section. Know that joining dissimilar metals requires adapters. Shut off water supply and drain plumbing lines fully. Cut out the damaged section using a hacksaw or pipe cutter. Remove old pipe; deburr and sand ends of remaining sections.

Dry fit the new replacement pipe piece to ensure proper length before permanent joining. Use couplings, clamps, or glue to attach new pipe ends to existing runs. For metal pipes like copper, lightly sand mating interior surfaces and apply joint compound before tightening couplings. Avoid over-tightening.

Plastic pipes like PVC require primer and cement to permanently bond joints. Apply both agents, then quickly push pipe into fittings while twisting to spread cement. Allow adhesive joints proper curing time before restoring water flow. Check for leaks as pressure increases. Insulate repair area to prevent refreezing. With care and patience, you can often replace deteriorated pipes yourself in accessible areas. But know your limits—significant re-piping projects warrant hiring a professional plumber.

Chapter 11

Maintaining and Improving Your Plumbing System

Routine plumbing maintenance keeps problems at bay before they disrupt your household. Let's review key preventative tasks DIYers should perform regularly.

Inspect visible pipes and fittings monthly for signs of drips, corrosion, cracks, or looseness. Tighten as needed and replace worn parts early.

Clear fixture drains regularly using baking soda and vinegar or a drain snake. Monthly drain cleaner treatment helps too. This prevents clog emergencies.

Check supply and drainage valves annually to ensure full closure and easy handle operation. Replace worn washers or faulty valves.

Flush hot water heaters yearly by draining and refilling the tank. This removes sediment that can accumulate at the bottom and reduce heating efficiency.

Test temperature and pressure relief valves on water heaters annually by briefly opening lever arms. This ensures they operate freely in an over pressure event.

Inspect toilet tank components for leaks and corrosion. Replace flappers, fill valves, and other components at first signs of trouble.

Caulk around sinks, tubs, and showers yearly. This prevents water intrusion under fixtures that can damage surround materials.

Upgrade old steel supply lines with modern flexible braided stainless-steel lines. This prevents burst failures.

Evaluate insulation on hot and cold water pipes. Add insulation where needed to reduce energy costs.

With proactive maintenance and early part replacement, plumbing systems will deliver reliable service year after year. An ounce of prevention is truly worth a pound of cure for household plumbing.

With concerns about water scarcity growing, improving household water efficiency is a smart goal. Many DIY upgrades reduce waste while saving on utility bills. Start by replacing any leaky faucets or malfunctioning toilets allowing phantom flushing. Leaks account for nearly 10% of indoor water waste. Stay on top of repairs. Install water-saving aerators on sink faucets and low-flow showerheads. Look for EPA WaterSense-certified models. Reduce flow without sacrificing performance. Upgrade old 3-5 gallon per flush toilets with new efficient 1.28 GPF models. Newer low-flow toilets flush well using much less water. Install dual-flush controls to save even more.

Insulate hot water pipes to reduce heat loss and wasted water as hot lines cool. Use foam pipe sleeves or fiberglass wraps to contain heat better. Repair drips immediately, as a slow drip can waste 20+ gallons per day, adding up over time. Don't tolerate any dripping faucets. Run full loads in dishwashers and washing machines. Partial loads waste electricity and water. Adjust load size settings to match your household needs. Adjust sprinklers and irrigation schedules monthly based on weather and lawn needs. Use moisture sensors to prevent unnecessary watering cycles. With smart upgrades and leak vigilance, households can reduce indoor and outdoor water usage substantially. Use resources wisely and save money in the process.

Replacing worn plumbing fixtures and appliances with modern, eco-friendly models improves daily function and efficiency. Let's look at some impactful green upgrades. For faucets, WaterSense labeled models restrict flow to 1.5 GPM without sacrificing performance. Touchless faucets also prevent waste by only running when needed. Showers using low-flow showerheads under 2.0 GPM get great pressure while using a fraction of the water. For baths, new tub spouts with integrated stops conserve.

High-efficiency toilets with a 1.28 GPF rating offer substantial water savings versus old 3-5 GPF models, while still flushing powerfully. Insulated electric tankless water heaters provide endless hot water on demand without standby energy losses. Their 99% efficiency saves energy.

ENERGY STAR certified dishwashers and washing machines save thousands of gallons yearly by optimizing wash cycles and load sensing. New instant hot water circulators reduce waste by delivering hot water to fixtures instantly without waiting. Recirculation systems work 24/7.

Finally, new touchless faucets and flushometers prevent phantom water loss from leaking valves. Auto-off stops flow after a set time. With smart upgrades, you can build a tightly integrated plumbing system that maximizes performance while minimizing resource waste. Invest in efficiency!

Upgrading aging plumbing components with modern materials improves efficiency, health, and reliability. Let's look at some impactful upgrades. Replacing old galvanized steel supply pipes with copper or flexible PEX lines removes corrosion risks and flow restrictions. PEX resists scale buildup too. Swap outdated grey polybutylene pipes prone to leaks with code-approved CPVC and PEX lines. Use resistant pipe materials for your water supply.

Replace lead service lines and solder with non-toxic alternatives. Avoid lead exposure from older supply components. Insulate exposed hot and cold pipes with pre-split foam sleeves to prevent energy loss and harmful condensation.

New push-fit fittings like ProPEX and Sharkbite simplify DIY connections without soldering or glue. They offer secure leakage-proof joints. Quiet noisy pipes by integrating flexible supply tubes or vibration dampening connection fittings like Sharkbite's Whisper. Install a whole house water filter system to remove contaminants, sediment, and odors for healthier water. Filters also protect appliances.

Finally, upgrade to a tankless or solar water heater for endless hot water with greater efficiency. Reduce energy use substantially. With thoughtful component upgrades, you can build a modern plumbing system optimized for safety, efficiency, and reliability. Invest wisely.

While many plumbing issues can be DIY projects, others require a licensed pro. Here are signs it's time to pick up the phone. Any leak, clog, or issue involving the main water supply line or municipal connection requires a plumber. Don't attempt to access these yourself. Major re-piping jobs, whole home pipe material upgrades, or re-piping under floors should also be left to the pros with proper tools and code knowledge.

For major leaks requiring significant drywall or floor demolition to trace the source, call a plumber to locate and repair it. Slow drains or clogged main sewer lines may need professional drain augering if DIY cable drain snakes can't clear the blockage after multiple attempts. Toilet clogs beyond the internal tank and bowl mechanisms may require professionally snaking the drain line from the stack or vent access point.

For significant leaks or corrosion on pipes in difficult to access wall and ceiling cavities, hire a pro willing to find and make repairs in tight spaces. If water heater repairs like element replacement involve plumbing beyond the average DIYer's skill level, hire a professional to ensure it's done properly. Know your limits! Correctly diagnosing issues like pressure loss or electrical faults requires true expertise. Don't cause more harm by guessing. Doing plumbing right takes experience. When projects get complex, call a trusted professional. They have the skills to solve problems you don't.

As we've explored, numerous opportunities exist to enhance water efficiency in your home's plumbing systems and daily habits. Taking advantage of these best practices is crucial for environmental and financial sustainability.

With climate change and population growth straining limited freshwater reserves globally, all of us must be conscious about our water use. Minor upgrades and simple behavior changes collectively have an outsized impact curbing millions of gallons of avoidable water waste.

The good news is that optimizing water efficiency is easy and affordable using the strategies detailed in this chapter. Improving your home's water performance empowers you to do your part for the planet while benefiting your wallet. Here are some key takeaways:

Detect even hairline leaks promptly using sound amplifiers and monitoring water meters. Fixing dripping faucets, showerheads and toilet valves eliminates thousands of gallons of leakage.

Install WaterSense labeled models when replacing toilets, faucets, showerheads and other fixtures. Advanced aerators, flow restrictors and dual-flush mechanisms guarantee performance.

Choose faucets with hands-free options or automatic shutoffs to prevent taps left running accidentally. Touchless activation also boosts hygiene.

Insulate all hot water pipes to minimize heat loss in delivery. Maintaining heat allows lowering water heater temperatures while retaining comfortable supply temperatures at fixtures.

Set water heater temperatures only as high as needed, around 120°F. This prevents scalding and reduces heat loss. Insulate tanks to reduce standby heating needed.

Take 5 minute showers or fewer. Install low flow showerheads allowing just 2 gallons per minute or less flow. Limiting shower length and water volume per use curbs waste.

Fill sink basins or buckets when washing dishes, produce, hands, etc rather than letting water run continuously. Reuse rinse water for cleaning or watering plants.

Run full loads in dishwashers and washing machines. Let faucets drip into empty machines to fill rather than pre-rinsing with full pressure flow. This exploits the cycle's water efficiency.

Install greywater recovery systems to reuse wastewater from appliances, fixtures and drains for irrigation, flushing, etc. Treated greywater offsets gallons pulled from limited drinking water sources.

Choose native, drought-tolerant plants that require minimal irrigation to maintain. Use soaker hoses or drip irrigation to precisely target roots instead of spraying. Water early mornings or at night to reduce evaporation.

If on private well water, monitor usage and aquifer levels. Avoid overtaxing groundwater faster than the surrounding environment can naturally recharge it.

Applying even a few of these impactful measures ensures your home plumbing operates at maximum efficiency. Small tweaks make a difference! Practicing conscientious water habits also motivates your community and future generations to value precious water resources.

With climate uncertainties ahead, becoming wise water stewards protects our environment along with our budgets. We hope this chapter has illuminated the path forward to achieve water efficiency gains through upgraded plumbing and mindful daily practices. What changes will you make today to become part of the solution? Our planet's intricate freshwater ecosystems will thank you!

79

Chapter 12

Advanced Plumbing Projects

When it's time to upgrade or replace your plumbing fixtures, installing new ones provides a great opportunity to improve your bathroom or kitchen's functionality and aesthetics. With some planning and proper technique, moderately skilled DIYers can handle installing new sinks, toilets, faucets, shower systems and other fixtures. Or you may choose to hire a professional plumber for a perfect finish.

Either way, installing new high-quality plumbing fixtures boosts your home's value and modernizes its systems for better convenience and performance. New water-efficient toilets and faucets conserve resources and save on utility bills. Premium finishes like granite, marble or stainless-steel update the look of kitchens and bathrooms. Added features like hand showers, temperature controls and touchless operation improve user experience.

The first step is disconnecting and extracting your outdated plumbing components. Turn off the water supply lines and drain the pipes. Disconnect the supply lines and drain connections. Remove any caulk, sealant or putty around the edges. Take out mounting screws or bolts to free the fixture. Carefully extract it without damaging the surrounding surfaces. Watch for falling debris or water spillage.

Thoroughly clean the empty space and repair any surface damage. Make sure the rough-in plumbing, like the stub-outs and drainpipe, are in good shape for the new fixture. Check for leaks or issues like misaligned pipes. Address any problems before moving on.

Read the manufacturer's instructions for your new sink, toilet, faucet, etc. Verify the necessary tools, materials and plumbing connections. Ensure the measurements and configuration will align with your existing rough-in plumbing. Purchase any missing supplies like fittings, hoses, valves or specialty tools.

Do a dry run to check the positioning of your fixture and plumbing hookups. Hold it in place, mark locations, and test orientations and alignments. For sinks, make sure you can reach the supply valves and drain connections. For toilets, verify there is enough space around the bowl and to the flange. Adjust your plans if needed before permanent installation.

When ready for full assembly, start by securing the new fixture in place. For sinks, set the basin into the countertop hole and use mounting clips below. For toilets, set wax ring seals and bolt down the bowl. Make all necessary drain connections, using new gaskets and fittings.
Attach the water supply lines, making sure to use the right type of tubing and adapters. Adjust supply stops and control valves. Secure any surrounding trim pieces or panels. Apply caulk around edges and wipe away excess. Remove any debris from drains and test for leaks. Finally, turn on the water supply and perform an initial test flush or flow. Enjoy your updated fixture!

Each type of plumbing fixture has unique installation steps, so follow manufacturer guidelines closely. Get help lifting very heavy components like cast iron tubs. Hire an electrician if electrical work is involved. Retain a carpenter for cabinetry modifications. Purchase upgraded supply lines like braided stainless steel. Add insulation around supply pipes if needed. Obtain any required plumbing permits and ensure compliance with codes. Do extensive planning for major projects like shower enclosures and tile work. Install isolation valves to facilitate future repairs. Keep instructions and parts for future maintenance needs. Proper installation ensures your new plumbing fixtures work flawlessly and aesthetically integrate into your redesigned space!

Remodeling a kitchen or bathroom offers an excellent opportunity to update worn fixtures, improve layout and storage, and install upgraded plumbing systems. Planning a remodel involves space planning, materials selection and integrating structural, electrical and plumbing changes. Having a professional plumber handle the plumbing portion ensures your new kitchen or bath works flawlessly for years to come.

Home Plumbing Mastery

When remodeling a kitchen, some popular plumbing projects include installing a polished chrome faucet with pull-out sprayer, an instant hot water dispenser, an undermount sink, air gaps for dishwashers, garbage disposals, and soft-closing water efficient faucets. Other options are potfiller faucets, hot/cold water dispensers, under-cabinet filtration systems, and touchless faucets.

For bathroom remodels, common plumbing upgrades are walk-in showers, steam units, jetted tubs, dual sink vanities, bidets, touchless faucets, and water efficient low-flow toilets. Claw foot tubs, rainfall showerheads, heated towel racks and freestanding tub filler faucets also create luxury experiences.

A remodel often involves reworking drain and vent lines to optimize layout and flow. Kitchen islands allow installing drainage underneath. New bathroom sink locations may require rerouting drains. Accessing pipes through walls and floors entails careful demolition and reconstruction. Plumbers have the expertise to determine optimal pipe sizing, gradients, trap configurations and venting to ensure high performance drainage free of leaks or sewer gas.

The plumber will connect all new sinks, faucets, toilets, tubs, showers and appliances to supply and drain lines. Gas stoves require specialized gas pipes and valves. For kitchens, they hook up dishwashers, refrigerators with ice makers, instant hot units and beverage dispensers. Installation may involve channeling pipe chases into cabinets and flooring. The right connections ensure adequate water flow and minimal pressure loss to each fixture.

New plumbing codes require water-conserving fixtures. Plumbers will customize options to your space like WaterSense dual-flush toilets, low-flow faucets and showerheads with flow restrictors. Leak proofing supplies enhances water efficiency. Proper drainage slope and venting optimizes water flow.

The plumber will help select fixtures and finishes that align with your kitchen or bath design aesthetic. Mixing metals, like brushed nickel faucets with oil-rubbed bronze cabinetry pulls, requires artistry. Tiling around tub surrounds necessitates careful measurements. Plumbers have an eye for symmetry, tub/shower alignments and creating elegant focal points with elements like freestanding tubs.

With careful planning and a skilled plumber handling the intricacies, your kitchen or bathroom remodel will result in a cohesive, high-performing space integrating top-notch form and advanced plumbing function. Be sure to get required permits and ensure all work complies with building codes.

Eventually, the plumbing in your home will require major renovation rather than just piecemeal repairs. Re-piping involves replacing old, corroded pipes with new ones to restore proper water supply and drainage. Since it affects your home's fundamental infrastructure, re-piping is best left to professional plumbers.

The decision to re-pipe results from a combination of factors like pipe lifespan, decline in water pressure and flow, leaks, high mineral/scale buildup, and drainage issues. Copper pipes typically need re-piping after 50-70 years. Galvanized steel only lasts 30-50 years before corroding. PVC can last 80+ years but tends to become brittle and crack over time.

How can you tell it's time for a full plumbing overhaul? Indications include:

Visible corrosion like green buildup in pipes or rusty water

Frequent clogs, leaks and bursts requiring repeated repairs

Reduced water pressure and sudden drops when using multiple fixtures

Banging or clanking sounds from water hammer in pipes

Higher utility bills from water loss through leaks

If your home is over 50 years old and you experience multiple symptoms, repiping may be your best option.

Re-piping involves removing old supply lines and drainpipes and replacing them with new ones. Plumbers will start by evaluating your current plumbing and inspecting for viability. They'll design a tailored plan addressing your home's needs and preferences.

During the work, plumbers will:

Shut off main water lines and drain the plumbing system

Remove flooring, walls and ceilings to access pipes

Extract old pipes and dispose properly

Replace corroded sewer lines if necessary

Install new copper, PEX or CPVC supply lines

Use advanced joining techniques for a seamless system

Position pipes to allow for optimal water flow

Reinstall fixtures like sinks, toilets and appliances

Seal openings or holes and test for leaks
Flush the system and confirm proper water pressure

While re-piping is a major undertaking, the benefits make it worthwhile:
Restored water pressure and flow
Elimination of leaks, bursts and water hammer
Prevention of further property damage
Reduced risk of mold, mildew and rot
Lower water and utility bills
Updated pipes that will last decades
Improved functionality of plumbing fixtures
Increased home value from updated infrastructure
When faced with chronic plumbing issues, re-piping is the gold standard solution that pays dividends through enhanced reliability, usage efficiency and longevity. Invest wisely—call professional plumbers to re-pipe your house and enjoy an optimally functioning system for years to come.

While water pipes constitute much of a home's plumbing, dealing with gas lines also falls under the plumber's purview. Gas lines power major appliances like water heaters, furnaces, stoves, ovens, fireplaces and outdoor grills. Plumbers are trained to work with natural gas infrastructure in a safe, code-compliant manner.
Home gas lines operate under low pressure and deliver gas from a main utility line to appliances via durable piping. Gas pipes are often steel, wrought iron or copper. The main shutoff valve allows gas flow to be quickly turned on or off. Meters track gas usage. Appliances have individual shutoff valves.
Gas line sizing depends on the BTU ratings of connected appliances and optimal gas flow. Undersized pipes can hinder appliance performance. Plumbers determine the right pipe diameters to allow sufficient volume without pressure drops.
Natural gas effectively powers appliances, but leaks can lead to dangerous explosions and fires. Warning signs include:
Rotten egg smell of added mercaptans
Hissing sound near appliances or pipes
Dead vegetation near gas lines
Flames lifting off burners or pilot lights
Increased moisture inside appliances
Higher and fluctuating gas bills

If you suspect a leak, evacuate immediately and call the gas company or fire department. Do not turn on/off appliances or lights or use phones until the area is declared safe.

Plumbers have specialized equipment to check for gas leaks. During inspections, they look for corrosion, loose fittings, cracks and missing pipe segments. Sophisticated gas detectors identify hard-to-find leaks.

For repairs, plumbers ensure proper pipe alignments and slopes maintain flow efficiency. They replace corroded sections and use code-approved fittings and pipes. Appliances get thoroughly cleaned, adjusted and tested. Plumbers verify safety valve operation, adequate ventilation and exhaust. All work aims to provide optimum gas conveyance while eliminating leakage risks.

Plumbers also install new gas lines during remodeling projects or when adding appliances like pool heaters and outdoor kitchens. New pipes get pressure tested and inspected before covering. For appliances, plumbers run dedicated gas lines and hook up fittings and valves. They customize extensions like flex gas lines for stoves and ovens. Sediment traps prevent blockages in crucial junctures.

For safety, plumbers emphasize durable piping, proper slopes, secure connections and adequate ventilation. They ensure all work meets stringent codes for gas systems. With their specialized skills, plumbers handle gas lines so homeowners can enjoy comfortable amenities without worrying about leaks, explosions or asphyxiation. Call certified professionals for any work involving natural gas plumbing.

Completing advanced plumbing projects requires in-depth skills, but also brings immense satisfaction. When done properly, complex installations stand the test of time and boost a home's functionality for years to come. If you carefully increased your knowledge through this chapter, you now grasp key lessons for plumbing success.

First, quality materials are essential. Don't cut corners to save pennies by using flimsy pipes or faulty valves that will fail quickly. Invest in sturdy fixture models from reputable brands that are built to last. Choose pipe types like copper and PEX that withstand pressure and corrosion. Spend a little more upfront and it will pay dividends for decades.

Equally important is mastering specialized plumbing techniques that meet codes and standards. From correctly sloping drains, to venting properly, to sealing joints tightly, to purging air and debris—every detail matters. There is little margin for error. Perfecting skills takes practice and patience. Hire experienced plumbers if complex projects intimidate you.

Safety always comes first too. Secure permits, disable electricity and water, provide ventilation and wear protective gear to prevent plumbing work accidents. Rushing through gas line or drain repairs to save time only heightens risks. Take things slow and steady.

Some tips to simplify success with advanced plumbing projects:

Educate yourself on building codes and instructions for your specific project. Knowledge is power! The more you understand proper procedures, the smoother the process will go.

Have the right tools and materials handy for each task. Readying your work area avoids delays rummaging for missing items mid-project.

Clearly communicate with plumbers or team members assisting you. Coordinate each step so everyone is on the same page.

Allow extra time in your schedule as unforeseen issues can arise. Don't let tight deadlines rush the work.

Get professional consultations if you're unsure how to approach any part of the job. It's never wrong to ask for guidance.

Document stages with photos as you go and keep detailed notes. This provides helpful records for future maintenance needs.

With preparation and perseverance, you can gain confidence tackling plumbing projects of any scale. Be willing to learn from mistakes. Each job completed adds to your skills. Soon you'll be the plumbing expert family and friends turn to for advice.

In closing, appreciate that few DIY pursuits are as rewarding as vanquishing a technically challenging plumbing project yourself or with assistance. Step back after the final wipe down and admire your handiwork. Those pipes now supply and safeguard the most valuable of necessities—water. Bask in your plumbing victory!

87

Chapter 13

Plumbing Maintenance and Prevention

Preventing problems is far easier than reacting to plumbing emergencies down the road. Developing consistent maintenance habits allows you to operate your plumbing proactively and catch minor issues before they become major.

Here are some key plumbing maintenance tasks DIYers should perform periodically:

Check exposed pipes and supply lines for leaks, corrosion, cracks and loose fittings. Look for green buildup, rust spots or mineral deposits indicating wear. Tighten any loose joints. Replace corroded sections. Wrap pipes with insulation if needed.

Over time, narrowing pipes can reduce water pressure. Test pressure at various fixtures. Sinks and tubs should get 20+ psi. Showers need 30+ psi. If pressure drops below these ranges, inspect your plumbing for blockages.

Ensure faucets, showers and appliances get adequate water flow. Low flow indicates obstructions or undersized pipes. Limescale buildup in showerheads and faucet aerators can restrict flow—disconnect and soak in vinegar to dissolve.

Turn valves on and off to verify proper operation and prevent them from seizing up. Spray lubricant if needed. Check sink and tub pop-up drain stoppers to ensure they engage fully.

Scan ceilings, under cabinets and basements for water stains indicating leaks. Confirm toilets aren't running constantly. Drips at faucets and spigots likely indicate worn washers/seals. Promptly fix any identified leaks.

Flushing water heaters annually removes mineral deposits that can accumulate in the tank. Turn off power. Attach a hose to the drain valve and route it to a floor drain. Open the valve and allow water to fully drain until it runs clear.

Inspect hoses behind refrigerators, dishwashers and washing machines. Replace any bulging, cracked or warped hoses which can burst and leak. Consider installing steel-reinforced hoses.

Minerals in water can collect on aerator screens on faucets. Unscrew the aerator and remove deposits. Clean drains screens in sinks, tubs and showers in the same manner. With periodic inspections and preventative maintenance, you can optimize your plumbing for maximum life, efficiency and trouble-free operation. Investing a little time upfront saves huge headaches later!

Depending on your climate, seasonal weather changes can wreak havoc on plumbing if you don't take preventative steps. From frigid winters to sweltering summers, temperature swings test the durability of pipes and fixtures. Advance preparation is key to avoiding seasonal plumbing disasters.

In regions with harsh winters, freezing temperatures can burst pipes and crack fixtures if water inside expands. To winterize, start by disconnecting and draining outdoor hoses. Insulate exposed exterior pipes, even those in crawlspaces and basements.

In a prolonged cold snap, keep cabinet doors open under sinks. Open indoor faucets slightly so pipes don't freeze. Prevent drafts near water meters. Consider installing heat tape on vulnerable pipes. Maintain optimal heating to keep indoor temperatures sufficiently warm.

Before trips, shut off main water lines and drainpipes, or maintain a trickle of warm water from faucets. Never turn your heat off completely while away in winter. Ask a house sitter to monitor temperature.

Finally, know where your main water shutoff is located so you can immediately stop water flow if pipes do freeze or burst. Learning to thaw pipes properly also proves useful. Never take plumbing for granted going into winter!

Just as cold brings unique issues, heat brings its own set of seasonal plumbing concerns. Prolonged hot weather can accelerate mineral deposit buildup and bacterial growth in stagnant water. Hoses, pipes and seals also become more prone to cracking.

In summer, monitor usage to check for spikes indicating leaks. Let taps drip to relieve pressure during peak hours. Flush systems thoroughly after periods of heavy use. Change filters more frequently when drawing higher volumes. Deep clean evaporative coolers and swamp coolers.

Check water heaters to ensure proper functioning—higher usage can reduce their lifespan. Inspect hoses and gaskets regularly in hot weather. Replace any showing wear. Avoid using harsh chemicals that could interact with higher water temperatures.

With proper seasonal preparations, you can avoid the headaches of frozen, burst or contaminated plumbing. Tailor your maintenance checklists to safeguard pipes, fixtures, appliances and water quality during both cold and hot weather extremes. A bit of foresight goes a long way toward uninterrupted year-round plumbing operations.

Few plumbing problems are as stressful (or messy) as drains suddenly overflowing with sewage. Clogged and backed-up drains disrupt household operations and can cause thousands in damage if left unchecked. Luckily, certain habits can minimize clogs.

Kitchen drains see loads of grease, food and soap residue. Fats and oils coats pipes, then food scraps stick and accumulate into blockages. Run hot water for 15 seconds while pouring grease into a jar rather than the sink. Use drain screens to catch food scraps.

Avoid pouring oils, grease, coffee grounds and eggshells down sinks. Compost food waste instead. Use a wire mesh drain cover to prevent larger scraps from entering pipes. Monthly, pour 1/2 cup baking soda and 1/2 cup vinegar down drains, waiting 10 minutes before rinsing with hot water.

Hair, soap residue and grime are the usual bathroom clog culprits. Prevent buildup with drain screens and hair catchers. Monthly, flush pipes with a mixture of 1/2 cup salt and 1/2 cup baking soda, followed by hot water. Avoid flushing sanitary products, cotton balls or dental floss which can snag and accumulate.

Lint traps on washing machines only catch so much—lint inevitably reaches pipes. Route discharge hoses to utility sinks rather than directly into branch drains. Use lint removal accessories on discharge lines. Avoid washing grease-stained rags which can solidify downstream.

If drains completely clog, first try a plunger to forcibly dislodge the blockage. Use a hand auger snake to fish out hair and gunk. Insert a shop vacuum hose firmly over drains and suck out debris. Caustic chemical drain cleaners dissolve organics and help clear mild clogs.

For severe clogs, call a professional plumber. Powerful motorized drain augers reach deep blockages and clear fully backed-up systems. Avoid amateur snaking which can damage pipes and worsen clogs.

With attentive habits and periodic maintenance, you can minimize phantom clog emergencies. But if disaster does strike, act quickly to clear blockages before overflows damage your home or belongings. Stay vigilant!

With regular maintenance, plumbing systems can operate reliably for decades. But there comes a point when age, wear and outdated technology make upgrading your plumbing vital. How do you know when it's time?

Here are indicators your home is due for plumbing upgrades:

Leaks and pipe bursts become frequent

Low water flow and pressure issues arise

Older galvanized steel pipes are rusting

Cast iron drainpipes are cracking

You want newer amenities like tankless water heaters, recirculation, or touchless faucets

Bathrooms and kitchen need extensive remodeling

Older fixtures are breaking down repeatedly

Essentially, if repair costs keep mounting, disruptions happen regularly, or you desire modern conveniences, it's time to upgrade.

With limited budgets, target upgrades delivering the most bang for buck. Replacing leaky polybutylene pipes and rusty galvanized supply lines should take priority over aesthetic upgrades. Re-piping provides huge benefits.

Next, upgrade bathrooms and kitchens used most frequently. Improvements like tankless water heaters, low-flow toilets and easy-access showers make everyday life more enjoyable.

Perform energy and water audits to identify savings opportunities. Upgrades like recirculation lines, insulated pipes and WaterSense fixtures optimize efficiency.

Finally, factor desired luxury features like steam showers and smart home monitoring into larger remodels. Integrate upgrades into repair projects for cost-effectiveness.

Compare the costs of frequent repairs and replacements against system upgrades. For example, replumbing a 1,500 square foot house with PEX may cost $4,000-8,000 but end recurring pipe ruptures.

Determine available rebates and tax credits which offset upgrade costs. Get professional estimates to understand your options. Though upgrades have higher initial costs, the investments pay dividends for years through enhanced performance, efficiency and reliability.

Through periodic upgrade projects, you can keep your plumbing aligned with your evolving needs and take advantage of the latest technology. View upgrades as long-term assets rather than short-term expenses. The improved performance justifies the investment.

After reviewing the many plumbing maintenance and prevention strategies in this chapter, we hope you feel empowered to take a proactive approach for your home's systems. Just like your own health benefits from good habits and preventative care, your plumbing needs the same regular TLC.

Consider establishing a routine maintenance schedule outlining specific tasks to perform seasonally, monthly and annually. Pop reminders in your calendar so upkeep never gets overlooked. Include details like ideal dates, steps for each activity, and parts/tools required.

Here are some maintenance quick wins with big impact:

Once a year, purge sediment from water heaters by flushing tanks. Inspect anodes and test pressure relief valves.

Check for leaks across all supply lines, drains, valves and fixtures twice yearly. One undetected leak can waste thousands of gallons over months.

Clear drains regularly using inexpensive plastic drain snakes monthly in problem areas prone to hair, soap scum and grease buildup.

Deep clean fixtures such as shower heads, faucet aerators and sink sprayers quarterly by descaling and removing trapped gunk.

Survey exposed water supply pipes in spring for external corrosion and insulation issues. Repair and insulate as needed before the next winter.

Test plumbing shutoff valves yearly to ensure they fully close. Replace any faulty valves that no longer completely stop water flow when closed.

Inspect exterior hose bibs and irrigation systems for freeze damage each spring. Repair any splits or cracks before summer watering starts.

Replace old steel braided supply lines with durable no-burst flex lines to avoid leaks as metal corrodes over decades of use.

Tour your home's foundation yearly looking for plumbing leaks evident by exterior staining. It's far better to seal foundation cracks before major pipe repairs are needed.

Catalogue model numbers of installed fixtures and tools. This allows easy replacement part identification when repairs are needed years down the road.

Prevention truly pays off by heading off serious issues before they materialize. Don't wait for the inevitable overflowing toilet or flooded basement. Conducting DIY walkthrough inspections makes you aware of plumbing health before disaster strikes.

Adopt the old maxim that an ounce of prevention is worth a pound of cure. Invest time upfront and maintain plumbing thoughtfully. Then relax and enjoy years of trouble-free operations! Your future self will thank you.

Chapter 14

Working Safely with Plumbing

While tackling minor repairs, remember plumbing can be hazardous if proper precautions aren't taken. From lung irritants to explosions, unaware DIYers risk serious bodily harm and property damage. Protect yourself with these safety measures.

Before any work, evaluate potential dangers like electric shock, lung damage, fire/explosion risks, and chemical contact hazards. Review the steps involved and safety gear needed. Have emergency numbers on hand.

De-energize systems before working on them. Turn off main water lines so pipes are depressurized before cutting or unscrewing joints. Shut off water heater power and gas lines. Doing so avoids electrocution, flooding, and explosions.

Wear safety glasses to prevent eye injuries when cutting pipe or drilling studs. Waterproof gloves protect against irritating fluxes and primers. Respirator masks filter out noxious pipe glue fumes. Knee pads provide comfort when crawling under sinks.

Clothing should fully cover arms and legs and have no dangling accessories that could catch. Sturdy non-slip shoes help avoid falls on ladders or wet floors. Have first-aid supplies nearby just in case.

Soldering, pipe glues and some cleaning chemicals release toxic vapors. Ensure adequate airflow to avoid nausea or lung damage. Open windows and use portable fans. Take breaks for fresh air. Unpleasant fumes indicate high levels of dangerous compounds.

Take precautions when handling primer, solvent, flux, sealant and cleaning products. Read labels and wear designated protective equipment. Dispose of chemicals properly rather than pouring down drains. Wash thoroughly after exposure. Call poison control if ingestion occurs.

Home Plumbing Mastery

By identifying risks and safeguards, you can perform plumbing repairs safely. Protection from lung damage, slips and serious cuts requires diligence but prevents severe injury and allows you to work confidently.

Having the right tools allows you to complete repairs correctly and safely. Improper use can damage plumbing, cause injuries, and leave issues unresolved. Follow these guidelines for using common plumbing tools.

Pipe wrenches like Stillsons tighten or loosen threaded pipes and fittings. The adjustable jaw locks onto pipe surfaces. Pull wrench handles in the proper direction. Don't over-torque. Use pipe joint compound to prevent seizing.

Channel locks tighten or loosen nuts, bolts and tubes. Grip round objects near the jaws rather than at the far handle ends for maximum leverage. Cover sharp ends with tape to prevent scratches.

Use hacksaws or PVC cutters to shorten pipes for removal or when joining segments. Keep blades perpendicular to cut straight. Lubricating blades reduces friction and overheating. Minimize burrs by finishing cuts with a metal file.

Bend small copper tubing by securing it in tube benders and slowly bringing handles together. Avoid tight bends or creasing. Bend rigid pipes like PVC using heat. Attempting room temperature bends can snap tubes. Make holes with carbide or diamond drill bits meant for the material. Use jigsaws to cut openings in walls and floors when installing new pipes. Hole saws cut clearance spaces. Let motors come to full speed before starting cuts.

Insert drain augers in sinks and tubs at proper angles to clear clogs. Turn augers clockwise when inserting then counterclockwise when withdrawing debris. Don't over-torque augers which can damage pipes or get tangled.

Avoid injuries by selecting the right tools for the task. Maintain sharp cutting edges. Replace damaged or dull tools that can lead to accidents.

Plumbing projects must adhere to local building codes for health, safety and functionality. Permits are required for most major work to ensure compliance. Understanding key code requirements helps DIYers avoid violations.

Most jurisdictions model codes after the International Plumbing Code (IPC) and International Residential Code (IRC). They dictate minimum fixtures per occupancy, pipe materials, venting rules, inspection requirements and more. The IPC covers large commercial buildings. IRC governs residential structures.

Both limit the distance drains can run horizontally without a vent. Certain plastic pipes are prohibited where hazards exist. Requirements for PEX vs copper differ. Knowing your local codes saves headaches during projects.

Permits are typically required for new fixtures, water heater swaps, re-piping, gas line work and major remodels. Permits apply even if you do the work yourself. Some minor repairs may be exempt. Checking your code will clarify.

Start the permit process by submitting an application with your scope of work, drawings and licensed plumber's involvement. Once approved, the permit allows scheduled inspections as work progresses. Passing inspections certify code compliance.

Permits protect homeowners by mandating proper materials, dimensions, venting and safe practices. Violations can jeopardize insurance claims and burden future buyers. Avoid headaches by permitting all covered projects. Homeowners doing their own plumbing must still follow codes. Consider taking applicable classes. Inspectors will expect workmanlike, quality execution. Some jurisdictions prohibit DIYers from permits involving gas lines.

Review codes related to your specific project. Keep approved materials and parts on hand. Have an experienced plumber double-check your plan and work. Build extra time and cost for multiple inspections. Permits make DIY jobs official.

While plumbing codes aim to protect public health and safety, they do require awareness during projects. Building departments help explain requirements.

In plumbing work, you may encounter hazardous waste requiring special disposal methods. Certain chemicals, ancient pipes, and appliance materials contain toxic substances. Improper disposal harms sanitation workers and the environment. Here are some hazardous materials you may encounter during DIY jobs.

Lead – Older pipes, solder and fixtures contained high lead levels. Inhaling or ingesting lead dust when removing these materials causes nervous system damage. Dispose at hazardous waste facilities.

Asbestos – Insulation on old pipes and water heaters contained cancer-causing asbestos fibers. Never saw or sand asbestos products. Hire abatement pros for removal.

Sediment – Mineral deposits inside old pipes and tanks contain heavy metals from corroded plumbing. Filter out sediment before drainage.

Caustic Chemicals - Drain cleaners, calcium/lime removers, and pipe cleaning chemicals irritate eyes, skin, and lungs. Follow label handling and dispose properly.

Mercury - Older thermostats and switches contained liquid mercury. If containers break, use gloves when cleaning. Dispose at hazardous waste centers.

Freon - Leaking refrigerant lines contain ozone-harming chemicals. Professional service safely captures and recycles coolant from damaged refrigerators.

Before any plumbing work, identify potential hazards involved. In most cases, disturbing risky materials requires professionals to contain, capture, neutralize or dispose of safely. Provide installers full details about your systems.

For chemicals or sediments you're handling yourself, utilize designated protective equipment. Follow usage, neutralizing, and disposal instructions carefully. Call local waste management authorities about proper disposal options. Dropping materials casually into normal trash or drains spreads toxins.

By identifying and properly addressing hazardous waste, everyone benefits—from your family, to waste workers, to the environment. Handling dangerous plumbing materials safely shows respect for human health and the planet. We hope this chapter has impressed upon you that plumbing projects demand safety-first practices. While it may seem excessive donning goggles and gloves just to fix a leaky pipe, taking proper precautions prevents severe and even fatal accidents. Respect the potential hazards involved.

Too often homeowners are in a rush and disregard safety protocols. They fail to turn off water and electricity before soldering. They neglect wearing particle masks when sawing pipes. They don't check for gas leaks before relighting appliances. Impatience leads to injuries and damage.

Follow all manufacturer directions and warning labels. Chemicals like drain cleaners and pipe fluxes can seriously harm if mishandled. Reporting new chemical odors in case gas lines are accidentally damaged during nearby work could prevent deadly explosions. Safety must be top priority.

Use common sense about risks of electrocution, toxic fumes, slippery surfaces, sharp edges and heavy objects. Assume serious danger until proven otherwise.

Confirm appliances like water heaters and HVAC systems are unplugged and lines are depressurized before any work nearby. Shut off main water lines for extensive plumbing repairs.

Wear protective eyewear, gloves, sturdy non-slip shoes, long sleeves and pants without cuffs that could catch. Masks guard lungs from fumes or airborne contaminants like lead dust.

Ventilate confined spaces and disperse any vapors, smoke or exhaust. Carbon monoxide is a silent killer. Confirm detectors are functioning.

Set up barricades and caution signs in the work zone. Rope off areas containing open trenches, ladders or dangerous tools.

Know the location of first-aid kits and fire extinguishers. Have your finger on the water shutoff valve in case swift action is needed.

Dispose of old pipes, fixtures and chemical containers properly rather than dumping hazardous waste. Children and pets snooping can get seriously sick.

Remain vigilant from start to finish on any plumbing job, no matter how mundane. Complacency causes accidents. Maintain your guard up, pay attention, use caution and please stay safe. The plumbing will still be there tomorrow if extra time is needed. No leaks or clogs warrant jeopardizing your health and home.

Chapter 15

Common Plumbing Myths and Mistakes

When confronted with clogged drains, many homeowners grab the strongest chemical cleaners in hopes of a quick fix. However, the dangers of caustic drain openers far outweigh any benefits. There are safer methods for clearing obstructions.

Alkaline, acidic, oxidizing and solvent drain cleaners can effectively dissolve organic material like hair, grease and soap scum. However, they present several hazards:

Corrosive to skin, eyes, lungs if spilled

React with other chemicals with toxic results

Damage pipes due to extreme pH

Release noxious fumes

Pollute groundwater if used excessively

Don't address mechanical obstructions

Common household ingredients provide drain-clearing power without the dangers above. A paste of baking soda and vinegar can break up debris and generate fizzing action. Boiling water helps liquefy fats and grease. Salt and baking soda loosen buildup when used regularly.

For deeper cleaning, try microbial drain cleaners containing bacteria or enzymes that digest organics. These won't harm pipes or humans. Be patient—natural cleaners work slower than corrosives.

The best solution is preventing clogs proactively through habits like:

Installing hair catchers

Minimizing grease down sinks

Flushing pipes monthly with hot water

Using strainers in showers and tubs

Avoid pouring harsh chemicals down drains. Handle any cleaners safely per directions and dispose properly. For stubborn clogs, call a plumber to mechanically snake drains rather than using risky caustic chemicals yourself.

While minor repairs like unclogging drains or replacing faucet washers are fine for DIYers, more complex plumbing jobs call for professional expertise. Attempting certain projects without enough skill risks safety hazards, code violations, and damage that ultimately costs more to fix.

Know your limits. Here are plumbing tasks best left to licensed plumbers: Sinks, toilets, tubs, and appliances require altering drain lines and water supply pipes. Proper venting and slope of drains takes professional know-how. Moving gas or electric lines is also tricky. Save yourself headaches and have plumbers install new major fixtures correctly the first time.

Re-piping an entire home or just the hot water system involves specialized skills. Plumbers determine the optimal pipe layout, material, sizing, fittings and joining techniques for a quality result. They can navigate tight spaces in walls, floors and attics that homeowners can't access.

Gas line installation and repair should only be done by certified professionals. Plumbers have skills like pipe sizing, perfect fittings, detecting leaks, pressure testing and ensuring compliance with codes. Mistakes with gas pipes, valves and vents can be catastrophic.

Kitchen and bathroom remodels require broad plumbing knowledge to integrate layout and design changes. This may include moving gas lines, adjusting drain positions, and complex installations best left to experts. Trying to DIY plumbing for aesthetics may look good but function poorly.

When any electrical work connects to plumbing, like with water heaters or appliance hookups, the general DIYer is at a disadvantage. Licensed electricians and plumbers have cross-training to ensure safe, code-compliant electrical integration.

Some regions prohibit any plumbing work without a license. Ensure you follow local regulations. When in doubt, hire a professional rather than risking disasters from flooded bathrooms, gas leaks, electrocution hazards or worse. Protect your family and home with qualified plumbing service.

With smart home tech invading plumbing, many fixtures now have electronic components that control functions. Touchless faucets, tankless heaters, bidets, and auto-flushing toilets promise high-tech convenience but also bring potential maintenance headaches.

On the plus side, electronic plumbing fixtures provide:
Advanced capabilities like voice activation, apps, remote monitoring
Premium features like heated seats, mood lighting, temperature precision
Improved hygiene and accessibility
Water conservation and energy savings

However, the downsides compared to basic fixtures include:
Higher initial cost
Complex repairs requiring specialists
Electromechanical glitches and malfunctions
Difficult troubleshooting diagnostic steps
Lack of manual override options
Planned obsolescence shortening lifespan

Before installing trendy smart plumbing, know that all electronics eventually fail. Weigh whether capabilities justify higher long-term service and replacement costs.

When glitches arise in electronic plumbing, DIY troubleshooting is limited. Start by rebooting devices or resetting WiFi connections as you would with smart home devices. Check for electrical and battery issues. Clean sensors like with touchless faucets.

For internal failures, retrieve any error codes if present. Consult installer and manufacturer troubleshooting guides. Beyond basic steps, repairs will require skilled professionals. Diagnosing bad circuit boards, servos and solid-state components in tankless heaters or auto-flush mechanisms is not DIY friendly.

If repairs exceed replacement cost, evaluate whether upgrading to another electronic fixture makes sense. Some older smart plumbing lacks connectivity to current home automation systems. As with consumer electronics, continued compatibility is not guaranteed.

Alternatively, replacing failed electronic fixtures with standard manual ones avoids repeat issues in the future. Though not as advanced, basic plumbing rarely malfunctions. Make choices balancing functionality, simplicity and lifespan. Avoid plumbing that may become e-waste in just a few years.

When plumbing issues arise, homeowners naturally aim for quick and inexpensive solutions. However, the temptation of budget materials and handyman hacks often backfires, resulting in safety risks and breakdowns needing proper repairs soon after.

Recognizing areas where spending a little more upfront prevents major future expenses and consequences is key to wise plumbing decisions. Here are prime examples:

Cheaper low-flow toilets tend to clog often, requiring repeated plunging and cleaning. Investing in a quality WaterSense approved high-efficiency toilet, though initially pricier, provides years of reliable service.

Homeowners sometimes try using cheap plastic PEX tubing and push-fit connectors without proper tools. Leaks result when connections fail. Hiring plumbers to install PEX correctly the first time is worth it.

Having a novice snake a tub drain with a flimsy plastic snake typically just worsens clogs when the snake breaks off inside pipes. Paying pros to thoroughly clear drains with industrial strength snakes prevents bigger pipe repairs.

Attempting installations sans permits seems like a cost saver but can lead to fines and code violations. Doing work up to code, even if it costs more, keeps you on the right side of inspectors.

Hiring cheap uncertified handymen for plumbing tasks like gas line repair risks disastrous errors. Always vet licenses since professional repairs by legit pros protect safety and compliance.

In all aspects of plumbing, spending a little extra for quality equipment installed properly by experts saves huge headaches, risks and costs over jerry-rigged DIY solutions. Think long-term and make wise investments that last.

This final chapter dispelling plumbing misconceptions concludes our journey together exploring home plumbing systems. By now we hope you feel equipped with accurate knowledge to make informed maintenance and repair decisions. Use your new understanding to avoid myths and mistakes that cause frustration, wasted time and money, damage, and injury risks.

Drain cleaners containing harsh chemicals like lye or acids may promise an easy fix for clogs but should be avoided. The caustic ingredients pose safety hazards if spilled and can gradually corrode pipes when poured regularly down drains. Homeowners drawn to drain cleaners for convenience often pay a high price later with extensive plumbing repairs needed from chemical damage. Instead, take preventive steps like using drain screens, minimizing grease sent to pipes, and flushing monthly with baking soda and vinegar. For stubborn clogs, call a professional plumber to mechanically snake drains rather than using dangerous chemicals yourself without proper training and protective gear.

It's tempting to roll up your sleeves and tackle complex plumbing projects like re-piping the entire home yourself as a do-it-yourselfer. However, advanced plumbing requires skills, tools and experience beyond what an average homeowner possesses. Projects involving extensive alterations, gas lines, drainage and venting are best left to licensed professionals rather than novices. Know your limits and don't let ego put your safety and home at risk. Make the wise choice hiring a reputable plumber for large scale changes to your home's vital underlying plumbing nervous system. Paying market rates for quality expertise protects your property and provides peace of mind.

High-tech "smart" plumbing fixtures with electronic controls, sensors, WiFi connectivity and cell phone apps may seem like exciting cutting-edge upgrades for bathrooms and kitchens. However, these complex components also introduce a higher risk of glitchy failures and malfunctions compared to traditional manual plumbing fixtures. Electronic sensor faucets, touchless toilets and tankless water heaters rely on circuit boards, motors and solenoids that the average homeowner cannot troubleshoot or repair when they unpredictably fail. Service calls to diagnose and change proprietary electronic parts often become frustrating and expensive. Before installing trendy high-tech plumbing, think carefully about increased long-term maintenance and replacement costs down the road balanced against any benefits like added convenience or accessibility. In most homes, choosing reliable and easily repairable manual fixtures may be the wiser choice over gadgetry that fails prematurely. Though not as exciting, traditional plumbing rarely suffers from the gremlins that frequently plague electronics. There is wisdom in keeping home systems simple rather than over-engineering solutions.

Home Plumbing Mastery

When faced with repairs, homeowners often balk at seemingly outrageous costs quoted by professional plumbers. However, parts, materials and fixtures involved are themselves pricey before even accounting for demanding physical labor and specialized skills and tools needed for installation and repairs. Plumbers must recoup their significant investments in ongoing training and licensing in order to provide customers with safe, high-quality work that meets codes. While DIY repairs may appear cheap on the surface, attempting plumbing without enough skill risks disaster through flooding, electrocution, gas leaks and sewer gases entering homes. Paying fair local market rates to have repairs done right the first time by qualified plumbers is truly an investment that protects the value and integrity of your home. Viewing plumbing experts as greedy or corrupt for charging commensurate prices for difficult but vital work reflects misunderstanding of the complexity involved. In most cases, you get what you pay for—cheap fixes by inexperienced handymen often translate to more headaches and costs not far down the road.

Our journey through the intricacies of home plumbing systems comes to an end, but hopefully your interest has just begun. With this book as your guide, you now have the knowledge to handle many common repairs yourself with confidence. More importantly, you've seen that And so we've reached the final page of our journey through the captivating world of home plumbing systems. I hope this book has equipped you with the knowledge and enthusiasm to handle common repairs and maintain your plumbing confidently. More importantly, I hope it reframed your perspective on plumbing itself.

It's easy to take plumbing for granted and see it as merely a practical concern for keeping your home livable. But I encourage you to appreciate plumbing for the marvel that it is—an engineering craft honed over millennia of human civilization. Your water supply and drainage systems represent the pinnacle of this long heritage.

The pipes in your home are an enduring legacy of human creativity, ingenuity, and perseverance. Countless innovators over thousands of years contributed incremental improvements that added up to the reliable plumbing we enjoy today. We stand on the shoulders of giants. I hope this book sparked a spirit of curiosity about the hidden wonders under your sinks and floors. Approach plumbing with a learner's mindset - always probing how things work and how you can work with them more skillfully. Even master plumbers remain students.

Share this curiosity by getting involved in the plumbing community, online and in person. Join forums, attend trade shows, and absorb the collective knowledge of decades. The world always needs more passionate plumbers. Above all, find satisfaction in a job well done. There's immense pride to be taken in spotting and solving a plumbing problem yourself. Don't shy away from challenges—let each one bolster your self-sufficiency. The pipes await.

Of course, I'm always here to help explain concepts if anything remains unclear! This concludes our journey for now, but there will always be more plumbing mysteries to unravel. Keep exploring. Keep learning. And most of all, keep plumbing!

plumbing consists of more than just leaky faucets and clogged drains. It's a complex infrastructure deeply integrated into our daily lives. One that protects our health, enables modern density, and provides creative challenges. Approach future plumbing projects with enthusiasm for mastering an endless craft. Find satisfaction in spotting and solving problems before they mushroom. And take pride that your skills keep your most essential home systems running smoothly. Yet don't feel compelled to tackle everything alone. Part of the plumbing ethos is community support and collaboration. Don't hesitate to consult fellow DIYers in online forums when stumped. And recognize when a task requires calling a professional.

Household plumbing may not seem glamorous. But it represents overcoming ancient struggles for comfort and safety through clever innovations. Your home's pipes are a direct link to history. Appreciate plumbing for the wonder it is. If this book provided nothing else, I hope it reframed your view of plumbing. Not as a dry technical topic, but as a vibrant craft with room for creativity. Your home awaits—go forth and make plumbing fun again! The pipes call for a new generation to carry the torch.

With this new fluency in plumbing concepts, you are equipped to make informed maintenance decisions, perform basic upkeep and repairs confidently, converse intelligently with professionals, and know when to seek expert assistance on major projects. View your home's pipes with fresh curiosity and appreciation.

Home Plumbing Mastery

While new do-it-yourself enthusiasts may feel overwhelmed initially when faced with the intricacies of plumbing, have faith that absorbing this knowledge at your own measured pace builds familiarity. Be willing to learn terminology and theory. Start slowly with basic handywork like changing fixture aerators or testing shut-off valves. As your competence grows, build up to more complex tasks. Follow tutorials to develop techniques. Consult plumbers for guidance when devising project plans. Invest in quality tools you will use repeatedly. Most importantly, know your limits and don't hesitate asking for help when prudent. Safety comes first.

In your journey from plumbing novice to capable homeowner, you now have the roadmap provided in this book. Our collaboration has equipped you to handle many repairs independently while empowering you to judge when professional services are needed. Your efforts keep money in your pocket and prevent catastrophic disasters like flooding. Take pride in your growing skills while remaining humble and exercising patience. Persevere through mistakes and remember: no leak or clog warrants compromising your health or home.

As a final note, consider passing along plumbing knowledge to the next generation. Teach your children about pipes and fixtures. Show them how to use tools properly and safely perform maintenance. Cultivate appreciation for this infrastructure we too often take for granted. Nurturing curiosity, competence and responsibility in younger generations ensures these skills get passed forward to perpetuate plumbing self-reliance.

Though mere pipes in the walls, home plumbing provides the lifeblood flowing through your house. We wish you smooth-running pipes bringing years of reliable service. Our waterways depend on mindful citizens valuing sustainability. May your stewardship help build a more water-aware world. Keep plumbing standards high and leaks few.